KRONECKER PRODUCTS & MATRIX CALCULUS WITH APPLICATIONS

KRONECKER PRODUCTS & MATRIX CALCULUS WITH APPLICATIONS

ALEXANDER GRAHAM
M.A., M.Sc., Ph.D.

Dover Publications
Garden City, New York

Bibliographical Note

This Dover edition, first published in 2018, is a slightly corrected republication of the work originally published by Ellis Horwood Limited, Chichester, West Sussex, England, in 1981.

International Standard Book Number

ISBN-13: 978-0-486-82417-8
ISBN-10: 0-486-82417-9

Manufactured in the United States of America
82417906
www.doverpublications.com

Table of Contents

Author's Preface

My purpose in writing this book is to bring to the attention of the reader, some recent developments in the field of Matrix Calculus. Although some concepts, such as Kronecker matrix products, the vector derivative etc. are mentioned in a few specialised books, no book, to my knowledge, is totally devoted to this subject. The interested researcher must consult numerous published papers to appreciate the scope of the concepts involved.

Matrix calculus applicable to square matrices was developed by Turnbull [29, 30] as far back as 1927. The theory presented in this book is based on the works of Dwyer and McPhail [15] published in 1948 and others mentioned in the Bibliography. It is more general than Turnbull's development and is applicable to non-square matrices. But even this more general theory has grave limitations, in particular it requires that in general the matrix elements are non constant and independent. A symmetric matrix, for example, is treated as a special case. Methods of overcoming some of these limitations have been suggested, but I am not aware of any published theory which is both quite general and simple enough to be useful.

The book is organised in the following way:

Chapter 1 concentrates on the preliminaries of matrix theory and notation which is found useful throughout the book. In particular, the simple and useful elementary matrix is defined. The vec operator is defined and many useful relations are developed. Chapter 2 introduces and establishes various important properties of the matrix Kronecker product.

Several applications of the Kronecker product are considered in Chapter 3. Chapter 4 introduces Matrix Calculus. Various derivatives of vectors are defined and the chain rule for vector differentiation is established. Rules for obtaining the derivative of a matrix with respect to one of its elements and conversely are discussed. Further developments in Matrix Calculus including derivatives of scalar functions of a matrix with respect to the matrix and matrix differentials are found in Chapter 5.

Chapter 6 deals with the derivative of a matrix with respect to a matrix.

This includes the derivation of expressions for the derivatives of both the matrix product and the Kronecker product of matrices with respect to a matrix. There is also the derivation of a chain rule of matrix differentiation. Various applications of at least some of the matrix calculus are discussed in Chapter 7.

By making use, whenever possible, of simple notation, including many worked examples to illustrate most of the important results and other examples at the end of each Chapter (except for Chapters 3 and 7) with solutions at the end of the book, I have attempted to bring a topic studied mainly at post-graduate and research level to an undergraduate level.

Symbols and Notation Used

$A, B, C \ldots$	matrices
A'	the transpose of A
a_{ij}	the (i,j)th element of the matrix A
$[a_{ij}]$	the matrix A having a_{ij} as its (i,j)th element
I_m	the unit matrix of order $m \times m$
$\mathbf{e}_i$	the unit vector
$\mathbf{e}$	the one vector (having all elements equal to one)
E_{ij}	the elementary matrix
0_m	the zero matrix of order $m \times m$
δ_{ij}	the Kronecker delta
$A_{\cdot i}$	the ith column of the matrix A
$A_{j\cdot}$	the jth row of A as a column vector
$A_{j\cdot}'$	the transpose of $A_{j\cdot}$ (a row vector)
$(A')_{\cdot i}$	the $i$the column of the matrix A'
$(A')_{\cdot i}'$	the transpose of the ith column of A' (that is, a row vector)
tr A	the trace of A
vec A	an ordered stock of columns of A
$A \otimes B$	the Kronecker product of A and B
iff	if and only if
diag $\{A\}$	the square matrix having elements $a_{11}, a_{22}, \ldots$ along its diagonal and zeros elsewhere
$\dfrac{\partial Y}{\partial x_{rs}}$	a matrix of the same order as Y
$\dfrac{\partial y_{ij}}{\partial X}$	a matrix of the same order as X
E_{rs}	an elementary matrix of the same order as X
E_{ij}	an elementary matrix of the same order as Y

KRONECKER PRODUCTS & MATRIX CALCULUS WITH APPLICATIONS

CHAPTER 1

Preliminaries

1.1 INTRODUCTION

In this chapter we introduce some notation and discuss some results which will be found very useful for the development of the theory of both Kronecker products and matrix differentiation. Our aim will be to make the notation as simple as possible although inevitably it will be complicated. Some simplification may be obtained at the expense of generality. For example, we may show that a result holds for a square matrix of order $n \times n$ and state that it holds in the more general case when A is of order $m \times n$. We will leave it to the interested reader to modify the proof for the more general case.

Further, we will often write

$$\sum_{i,j} a_{ij} \quad \text{or} \quad \sum_i \sum_j a_{ij} \quad \text{or just} \quad \sum \sum a_{ij} \quad \text{instead of} \quad \sum_{i=1}^{m} \sum_{j=1}^{n} a_{ij},$$

when the summation limits are obvious from the context.

Many other simplifications will be used as the opportunities arise. Unless of particular importance, we shall not state the order of the matrices considered. It will be assumed that, for example, when taking the product AB or ABC the matrices are conformable.

1.2 UNIT VECTORS AND ELEMENTARY MATRICES

The **unit vectors** of order n are defined as

$$\mathbf{e}_1 = \begin{bmatrix} 1 \\ 0 \\ 0 \\ \vdots \\ 0 \end{bmatrix}, \ \mathbf{e}_2 = \begin{bmatrix} 0 \\ 1 \\ 0 \\ \vdots \\ 0 \end{bmatrix}, \ \ldots, \ \mathbf{e}_n = \begin{bmatrix} 0 \\ 0 \\ 0 \\ \vdots \\ 1 \end{bmatrix} \tag{1.1}$$

The **one vector** of order n is defined as

$$\mathbf{e} = \begin{bmatrix} 1 \\ 1 \\ 1 \\ \vdots \\ 1 \end{bmatrix} \tag{1.2}$$

From (1.1) and (1.2), obtain the relation

$$\mathbf{e} = \Sigma \mathbf{e}_i \tag{1.3}$$

The **elementary matrix** E_{ij} is defined as the matrix (of order $m \times n$) which has a unity in the (i,j)th position and all other elements are zero.

For example,

$$E_{23} = \begin{bmatrix} 0 & 0 & 0 & \dots & 0 \\ 0 & 0 & 1 & \dots & 0 \\ 0 & 0 & 0 & \dots & 0 \\ \vdots & & & & \\ 0 & 0 & 0 & \dots & 0 \end{bmatrix} \tag{1.4}$$

The relation between $\mathbf{e}_i$, $\mathbf{e}_j$ and E_{ij} is as follows

$$E_{ij} = \mathbf{e}_i \mathbf{e}_j' \tag{1.5}$$

where $\mathbf{e}_j'$ denotes the transposed vector (that is, the row vector) of $\mathbf{e}_j$.

Example 1.1

Using the unit vectors of order 3

(i) form E_{11}, E_{21}, and E_{23}

(ii) write the unit matrix of order 3×3 as a sum of the elementary matrices.

Solution

(i)

$$E_{11} = \begin{bmatrix} 1 \\ 0 \\ 0 \end{bmatrix} [1 \;\; 0 \;\; 0] = \begin{bmatrix} 1 & 0 & 0 \\ 0 & 0 & 0 \\ 0 & 0 & 0 \end{bmatrix}$$

$$E_{21} = \begin{bmatrix} 0 \\ 1 \\ 0 \end{bmatrix} [1 \;\; 0 \;\; 0] = \begin{bmatrix} 0 & 0 & 0 \\ 1 & 0 & 0 \\ 0 & 0 & 0 \end{bmatrix}$$

$$E_{23} = \begin{bmatrix} 0 \\ 1 \\ 0 \end{bmatrix} [0 \;\; 0 \;\; 1] = \begin{bmatrix} 0 & 0 & 0 \\ 0 & 0 & 1 \\ 0 & 0 & 0 \end{bmatrix}$$

(ii) $$I = E_{11} + E_{22} + E_{33} = \sum_{i=1}^{3} \mathbf{e}_i \mathbf{e}_i' .$$

The **Kronecker delta** δ_{ij} is defined as

$$\delta_{i,j} = \begin{cases} 1 & \text{if } i = j \\ 0 & \text{if } i \neq j \end{cases}$$

it can be expressed as

$$\delta_{ij} = \mathbf{e}_i'\mathbf{e}_j = \mathbf{e}_j'\mathbf{e}_i \quad . \tag{1.6}$$

We can now determine some relations between unit vectors and elementary matrices.

$$\begin{aligned} E_{ij}\mathbf{e}_r &= \mathbf{e}_i\mathbf{e}_j'\mathbf{e}_r \quad \text{(by 1.5)} \\ &= \delta_{jr}\mathbf{e}_i \end{aligned} \tag{1.7}$$

and

$$\begin{aligned} \mathbf{e}_r'E_{ij} &= \mathbf{e}_r'\mathbf{e}_i\mathbf{e}_j' \\ &= \delta_{ri}\mathbf{e}_j' \quad . \end{aligned} \tag{1.8}$$

Also

$$E_{ij}E_{rs} = \mathbf{e}_i\mathbf{e}_j'\mathbf{e}_r\mathbf{e}_s' = \delta_{jr}\mathbf{e}_i\mathbf{e}_s' = \delta_{jr}E_{is} \quad . \tag{1.9}$$

In particular if $r = j$, we have

$$E_{ij}E_{js} = \delta_{jj}E_{is} = E_{is}$$

and more generally

$$E_{ij}E_{js}E_{sm} = E_{is}E_{sm} = E_{im} \quad . \tag{1.10}$$

Notice from (1.9) that

$$E_{ij}E_{rs} = 0 \text{ if } j \neq r .$$

1.3 DECOMPOSITIONS OF A MATRIX

We consider a matrix A of order $m \times n$ having the following form

$$A = \begin{bmatrix} a_{11} & a_{12} & \dots & a_{1n} \\ a_{21} & a_{22} & \dots & a_{2n} \\ \vdots & & & \\ a_{m1} & a_{m2} & \dots & a_{mn} \end{bmatrix} = [a_{ij}] \tag{1.11}$$

We denote the n **columns** of A by $A_{\cdot 1}, A_{\cdot 2}, \dots A_{\cdot n}$. So that

$$A_{\cdot j} = \begin{bmatrix} a_{1j} \\ a_{2j} \\ \vdots \\ a_{mj} \end{bmatrix} \quad (j = 1, 2, \dots, n) \tag{1.12}$$

and the m rows of A by $A_{1\cdot}, A_{\cdot 2}, \ldots A_{m\cdot}$ so that

$$A_{i\cdot} = \begin{bmatrix} a_{i1} \\ a_{i2} \\ \vdots \\ a_{in} \end{bmatrix} \quad (i = 1, 2, \ldots, m) \tag{1.13}$$

Both the $A_{\cdot j}$ and the $A_{i\cdot}$ are **column vectors**. In this notation we can write A as the (partitioned) matrix

$$A = [A_{\cdot 1}\, A_{\cdot 2} \ldots A_{\cdot n}] \tag{1.14}$$

or as

$$A = [A_{1\cdot}\, A_{2\cdot} \ldots A_{m\cdot}]' \tag{1.15}$$

(where the prime means 'the transpose of').

For example, let

$$A = \begin{bmatrix} a_{11} & a_{12} \\ a_{21} & a_{22} \end{bmatrix}$$

so that

$$A_{1\cdot} = \begin{bmatrix} a_{11} \\ a_{12} \end{bmatrix} \quad \text{and} \quad A_{2\cdot} = \begin{bmatrix} a_{21} \\ a_{22} \end{bmatrix}$$

then

$$[A_{1\cdot}A_{2\cdot}]' = \begin{bmatrix} a_{11} & a_{21} \\ a_{12} & a_{22} \end{bmatrix}' = \begin{bmatrix} a_{11} & a_{12} \\ a_{21} & a_{22} \end{bmatrix} = A\ .$$

The elements, the columns and the rows of A can be expressed in terms of the unit vectors as follows:

$$\text{The } j\text{th column } A_{\cdot j} = A\mathbf{e}_j \tag{1.16}$$

$$\text{The } i\text{th row } A_{i\cdot}' = \mathbf{e}_i' A. \tag{1.17}$$

So that

$$A_{i\cdot} = (\mathbf{e}_i' A)' = A'\mathbf{e}_i\ . \tag{1.18}$$

The (i,j)th element of A can now be written as

$$a_{ij} = \mathbf{e}_i' A \mathbf{e}_j = \mathbf{e}_j' A' \mathbf{e}_i\ . \tag{1.19}$$

We can express A as the sum

$$A = \Sigma\Sigma a_{ij} E_{ij} \tag{1.20}$$

(where the E_{ij} are of course of the same order as A) so that

$$A = \sum_i \sum_j a_{ij} \mathbf{e}_i \mathbf{e}_j'\ . \tag{1.21}$$

From (1.16) and (1.21)

$$
\begin{aligned}
A_{\cdot j} = A\mathbf{e}_j &= \left(\sum_i\sum_j a_{ij}\mathbf{e}_i\mathbf{e}_j'\right)\mathbf{e}_j \\
&= \sum_i\sum_j a_{ij}\mathbf{e}_i(\mathbf{e}_j'\mathbf{e}_j) \\
&= \sum_i a_{ij}\mathbf{e}_i \,. \qquad (1.22)
\end{aligned}
$$

Similarly

$$A_{i\cdot} = \sum_j a_{ij}\mathbf{e}_j \qquad (1.23)$$

so that

$$A_{i\cdot}' = \sum_j a_{ij}\mathbf{e}_j' \,. \qquad (1.24)$$

It follows from (1.21), (1.22), and (1.24) that

$$A = \Sigma A_{\cdot j}\mathbf{e}_j' \qquad (1.25)$$

and

$$A = \Sigma \mathbf{e}_i A_{i\cdot}' \,. \qquad (1.26)$$

Example 1.2

Write the matrix

$$A = \begin{bmatrix} a_{11} & a_{12} \\ a_{21} & a_{22} \end{bmatrix}$$

as a sum of: (i) column vectors of A; (ii) row vectors of A.

Solutions

(i) Using (1.25)

$$
\begin{aligned}
A &= A_{\cdot 1}\mathbf{e}_1' + A_{\cdot 2}\mathbf{e}_2' \\
&= \begin{bmatrix} a_{11} \\ a_{21} \end{bmatrix} [1 \quad 0] + \begin{bmatrix} a_{12} \\ a_{22} \end{bmatrix} [0 \quad 1]
\end{aligned}
$$

Using (1.26)

$$
\begin{aligned}
A &= \mathbf{e}_1 A_{1\cdot}' + \mathbf{e}_2 A_{2\cdot}' \\
&= \begin{bmatrix} 1 \\ 0 \end{bmatrix} [a_{11} \quad a_{12}] + \begin{bmatrix} 0 \\ 1 \end{bmatrix} [a_{21} \quad a_{22}] \,.
\end{aligned}
$$

There exist interesting relations involving the elementary matrices operating on the matrix A.

For example

$$
\begin{aligned}
E_{ij}A &= \mathbf{e}_i\mathbf{e}_j'A \qquad \text{(by 1.5)} \\
&= \mathbf{e}_i A_{j\cdot}' \qquad \text{(by 1.17)} \qquad (1.27)
\end{aligned}
$$

Similarly

$$AE_{ij} = A\mathbf{e}_i\mathbf{e}_j' = A._i\mathbf{e}_j' \quad \text{(by 1.16)} \tag{1.28}$$

so that

$$AE_{jj} = A._j\mathbf{e}_j' \tag{1.29}$$

$$AE_{ij}B = A\mathbf{e}_i\mathbf{e}_j'B = A._iB_j.' \quad \text{(by 1.28 and 1.27)} \tag{1.30}$$

$$\begin{aligned} E_{ij}AE_{rs} &= \mathbf{e}_i\mathbf{e}_j'A\mathbf{e}_r\mathbf{e}_s' && \text{(by 1.5)} \\ &= \mathbf{e}_i a_{jr}\mathbf{e}_s' && \text{(by 1.19)} \\ &= a_{jr}\mathbf{e}_i\mathbf{e}_s' = a_{jr}E_{is} \end{aligned} \tag{1.31}$$

In particular

$$E_{jj}AE_{rr} = a_{jr}E_{jr} \tag{1.32}$$

Example 1.3

Use elementary matrices and/or unit vectors to find an expression for

(i) The product AB of the matrices $A = [a_{ij}]$ and $B = [b_{ij}]$.

(ii) The kth column of the product AB

(iii) The kth column of the product XYZ of the matrices $X = [x_{ij}]$, $Y = [y_{ij}]$ and $Z = [z_{ij}]$

Solutions

(i) By (1.25) and (1.29)

$$A = \Sigma A._j\mathbf{e}_j' = \Sigma AE_{jj}\,,$$

hence

$$\begin{aligned} AB &= \Sigma(AE_{jj})B = \Sigma(A\mathbf{e}_j)(\mathbf{e}_j'B) \\ &= \Sigma A._jB_j.' \quad \text{(by (1.16) and (1.17)} \end{aligned}$$

(ii) (a)

$$(AB)._k = (AB)\mathbf{e}_k = A(B\mathbf{e}_k) = AB._k \quad \text{by (1.16)}$$

(b) From (i) above we can write

$$\begin{aligned} (AB)._k &= \sum_j(A\mathbf{e}_j\mathbf{e}_j'B)\mathbf{e}_k = \sum_j(A\mathbf{e}_j)(\mathbf{e}_j'B\mathbf{e}_k) \\ &= \sum_j A._jb_{jk} \quad \text{by (1.16) and (1.19)}\ . \end{aligned}$$

(iii)

$$\begin{aligned} (XYZ)._k &= \sum_j z_{jk}(XY)._j \quad \text{by (ii)(b) above} \\ &= \sum_j (z_{jk}X)Y._j \quad \text{by (ii)(a) above.} \end{aligned}$$

1.4 THE TRACE FUNCTION

The **trace** (or the **spur**) of a square matrix A of order $(n \times n)$ is the sum of the diagonal terms

$$\sum_{i=1}^{n} a_{ii}\ .$$

We write

$$\operatorname{tr} A = \Sigma a_{ii} \; . \tag{1.33}$$

From (1.19) we have

$$a_{ii} = \mathbf{e}_i' A \mathbf{e}_i \, ,$$

so that

$$\operatorname{tr} A = \Sigma \mathbf{e}_i' A \mathbf{e}_i \; . \tag{1.34}$$

From (1.16) and (1.34) we find

$$\operatorname{tr} A = \Sigma \mathbf{e}_i' A_{\cdot i} \tag{1.35}$$

and from (1.17) and (1.34)

$$\operatorname{tr} A = \Sigma A_{i\cdot}' \mathbf{e}_i' \; . \tag{1.36}$$

We can obtain similar expression for the trace of a product AB of matrices.

For example

$$\operatorname{tr} AB = \sum_i \mathbf{e}_i' AB \mathbf{e}_i \tag{1.37}$$

$$= \sum_j \sum_i (\mathbf{e}_i' A \mathbf{e}_j)(\mathbf{e}_j' B \mathbf{e}_i) \qquad \text{(See Ex. 1.3)}$$

$$= \sum_j \sum_i a_{ij} b_{ji} \tag{1.38}$$

Similarly

$$\operatorname{tr} BA = \sum_j \mathbf{e}_j' BA \mathbf{e}_j$$

$$= \sum_i \sum_j (\mathbf{e}_j' B \mathbf{e}_i)(\mathbf{e}_i' A \mathbf{e}_j)$$

$$= \sum_i \sum_j b_{ji} a_{ij} \; . \tag{1.39}$$

From (1.38) and (1.39) we find that

$$\operatorname{tr} AB = \operatorname{tr} BA \; . \tag{1.40}$$

From (1.16), (1.17) and (1.37) we have

$$\operatorname{tr} AB = \Sigma A_{i\cdot}' B_{\cdot i} \; . \tag{1.41}$$

Also from (1.40) and (1.41)

$$\operatorname{tr} AB = \Sigma B_{i\cdot}' A_{\cdot i} \; . \tag{1.42}$$

Similarly

$$\operatorname{tr} AB' = \Sigma A_{i\cdot}' B_i \; . \tag{1.43}$$

and since $\operatorname{tr} AB' = \operatorname{tr} A'B$

$$\operatorname{tr} AB' = \Sigma A_{\cdot i}' B_{\cdot i} \; . \tag{1.44}$$

Two important properties of the trace are

$$\operatorname{tr}(A+B) = \operatorname{tr} A + \operatorname{tr} B \tag{1.45}$$

and

$$\operatorname{tr}(\alpha A) = \alpha \operatorname{tr} A \tag{1.46}$$

where α is a scalar.

These properties show that trace is a **linear** function.

For real matrices A and B the various properties of $\operatorname{tr}(AB')$ indicated above show that it is an inner product and is sometimes written as

$$\operatorname{tr}(AB') = (A, B)$$

1.5 THE VEC OPERATOR

We shall make use of a vector valued function denoted by vec A of a matrix A defined by Neudecker [22].

If A is of order $m \times n$

$$\operatorname{vec} A = \begin{bmatrix} A_{\cdot 1} \\ A_{\cdot 2} \\ \vdots \\ A_{\cdot n} \end{bmatrix} . \tag{1.47}$$

From the definition it is clear that vec A is a vector of order mn.

For example if

$$A = \begin{bmatrix} a_{11} & a_{12} \\ a_{21} & a_{22} \end{bmatrix}$$

then

$$\operatorname{vec} A = \begin{bmatrix} a_{11} \\ a_{21} \\ a_{12} \\ a_{22} \end{bmatrix} .$$

Example 1.4

Show that we can write $\operatorname{tr} AB$ as $(\operatorname{vec} A')' \operatorname{vec} B$

Solution

By (1.37)

$$\begin{aligned} \operatorname{tr} AB &= \sum_i \mathbf{e}_i' AB\mathbf{e}_i \\ &= \sum_i A_{i\cdot}' B_{\cdot i} \qquad \text{by (1.16) and (1.17)} \\ &= \sum_i (A')_{\cdot i}' B_{\cdot i} \end{aligned}$$

(since the ith row of A is the ith column of A')

Hence (assuming A and B of order $n \times n$)

$$\operatorname{tr} AB = [(A')_{.1}{}'(A')_{.2}{}' \ldots (A')_{.n}{}'] \begin{bmatrix} B_{.1} \\ B_{.2} \\ \vdots \\ B_{.n} \end{bmatrix}$$

$$= (\operatorname{vec} A')' \operatorname{vec} B$$

Before discussing a useful application of the above we must first agree on notation for the transpose of an elementary matrix, we do this with the aid of an example.

$$\text{Let } X = \begin{bmatrix} x_{11} & x_{12} & x_{13} \\ x_{21} & x_{22} & x_{23} \end{bmatrix},$$

then an elementary matrix associated with will X will also be of order (2×3).

For example, one such matrix is

$$E_{12} = \begin{bmatrix} 0 & 1 & 0 \\ 0 & 0 & 0 \end{bmatrix}.$$

The transpose of E_{12} is the matrix

$$E'_{12} = \begin{bmatrix} 0 & 0 \\ 1 & 0 \\ 0 & 0 \end{bmatrix}.$$

Although at first sight this notation for the transpose is sensible and is used frequently in this book, there are associated snags. The difficulty arises when the suffix notation is not only indicative of the matrix involved but also determines specific elements as in equations (1.31) and (1.32). On such occasions it will be necessary to use a more accurate notation indicating the matrix order and the element involved. Then instead of E_{12} we will write $E_{12}(2 \times 3)$ and instead of E'_{12} we write $E_{21}(3 \times 2)$.

More generally if X is a matrix or order $(m \times n)$ then the transpose of

$$E_{rs}(m \times n)$$

will be written as

$$E'_{rs}$$

unless an accurate description is necessary, in which case the transpose will be written as

$$E_{sr}(n \times m) .$$

Now for the application of the result of Example 1.4 which will be used later on in the book.

From the above

$$\begin{aligned} \operatorname{tr} E'_{rs}A &= (\operatorname{vec} E_{rs})' (\operatorname{vec} A) \\ &= a_{rs} \end{aligned}$$

where a_{rs} is the (r, s)th element of the matrix A.

We can of course prove this important result by a more direct method.

$$\begin{aligned} \operatorname{tr} E'_{rs}A &= \sum_k \mathbf{e}'_k E'_{rs} A \mathbf{e}_k \\ &= \sum_{i,j,k} a_{ij} \mathbf{e}'_k \mathbf{e}_s \mathbf{e}'_r \mathbf{e}_i \mathbf{e}'_j \mathbf{e}_k \qquad \left(\text{since } A = \sum_{i,j} a_{ij} E_{ij}\right) \\ &= \sum_{i,j,k} a_{ij} \delta_{ks} \delta_{ri} \delta_{jk} = a_{rs} \end{aligned}$$

Problems for Chapter 1

(1) The matrix A is of order $(4 \times n)$ and the matrix B is of order $(n \times 3)$. Write the product AB in terms of the rows of A, that is, $A_{1\cdot}, A_{2\cdot}, \ldots$ and the columns of B, that is, $B_{\cdot 1}, B_{\cdot 2}, \ldots$.

(2) Describe in words the matrices

(a) AE_{ik} and (b) $E_{ik}A$.

Write these matrices in terms of an appropriate product of a row or a column of A and a unit vector.

(3) Show that

(a) $\operatorname{tr} ABC = \sum_i A'_{i\cdot} BC_{\cdot i}$

(b) $\operatorname{tr} ABC = \operatorname{tr} BCA = \operatorname{tr} CAB$

(4) Show that $\operatorname{tr} AE_{ij} = a_{ji}$

(5) $B = [b_{ij}]$ is a matrix of order $(n \times n)$
$\operatorname{diag}\{B\} = \operatorname{diag}\{b_{11}, b_{22}, \ldots, b_{nn}\} = \sum b_{ii} E_{ii}$.
Show that if

$$a_{ij} = \operatorname{tr} BE_{ij} \delta_{ij}$$

then $A = [a_{ij}] = \operatorname{diag}\{B\}$.

CHAPTER 2

The Kronecker Product

2.1 INTRODUCTION

Kronecker product, also known as **a direct product** or **a tensor product** is a concept having its origin in group theory and has important applications in particle physics. But the technique has been successfully applied in various fields of matrix theory, for example in the solution of matrix equations which arise when using Lyapunov's approach to the stability theory. The development of the technique in this chapter will be as a topic within the scope of matrix algebra.

2.2 DEFINITION OF THE KRONECKER PRODUCT

Consider a matrix $A = [a_{ij}]$ of order $(m \times n)$ and a matrix $B = [b_{ij}]$ of order $(r \times s)$. The Kronecker product of the two matrices, denoted by $A \otimes B$ is defined as the partitioned matrix

$$A \otimes B = \begin{bmatrix} a_{11}B & a_{12}B & \dots & a_{1n}B \\ a_{21}B & a_{22}B & \dots & a_{2n}B \\ \vdots & \vdots & & \vdots \\ a_{m1}B & a_{m2}B & \dots & a_{mn}B \end{bmatrix} \tag{2.1}$$

$A \otimes B$ is seen to be a matrix of order $(mr \times ns)$. It has mn blocks, the (i,j)th block is the matrix $a_{ij}B$ of order $(r \times s)$.

For example, let

$$A = \begin{bmatrix} a_{11} & a_{12} \\ a_{21} & a_{22} \end{bmatrix}, \quad B = \begin{bmatrix} b_{11} & b_{12} \\ b_{21} & b_{22} \end{bmatrix},$$

then

$$A \otimes B = \begin{bmatrix} a_{11}B & a_{12}B \\ a_{21}B & a_{22}B \end{bmatrix} = \begin{bmatrix} a_{11}b_{11} & a_{11}b_{12} & a_{12}b_{11} & a_{12}b_{12} \\ a_{11}b_{21} & a_{11}b_{22} & a_{12}b_{21} & a_{12}b_{22} \\ a_{21}b_{11} & a_{21}b_{12} & a_{22}b_{11} & a_{22}b_{12} \\ a_{21}b_{21} & a_{21}b_{22} & a_{22}b_{21} & a_{22}b_{22} \end{bmatrix}.$$

Notice that the Kronecker product is defined irrespective of the order of the matrices involved. From this point of view it is a more general concept than matrix multiplication. As we develop the theory we will note other results which are more general than the corresponding ones for matrix multiplication.

The Kronecker product arises naturally in the following way. Consider two linear transformations

$$\mathbf{x} = A\mathbf{z} \quad \text{and} \quad \mathbf{y} = B\mathbf{w}$$

which, in the simplest case take the form

$$\begin{bmatrix} x_1 \\ x_2 \end{bmatrix} = \begin{bmatrix} a_{11} & a_{12} \\ a_{21} & a_{22} \end{bmatrix} \begin{bmatrix} z_1 \\ z_2 \end{bmatrix} \quad \text{and} \quad \begin{bmatrix} y_1 \\ y_2 \end{bmatrix} = \begin{bmatrix} b_{11} & b_{12} \\ b_{21} & b_{22} \end{bmatrix} \begin{bmatrix} w_1 \\ w_2 \end{bmatrix} . \tag{2.2}$$

We can consider the two transformations simultaneously by defining the following vectors

$$\boldsymbol{\mu} = \mathbf{x} \otimes \mathbf{y} = \begin{bmatrix} x_1y_1 \\ x_1y_2 \\ x_2y_1 \\ x_2y_2 \end{bmatrix} \quad \text{and} \quad \mathbf{v} = \mathbf{z} \otimes \mathbf{w} = \begin{bmatrix} z_1w_1 \\ z_1w_2 \\ z_2w_1 \\ z_2w_2 \end{bmatrix} . \tag{2.3}$$

To find the transformation between $\boldsymbol{\mu}$ and $\mathbf{v}$, we determine the relations between the components of the two vectors.

For example,

$$\begin{aligned} x_1y_1 &= (a_{11}z_1 + a_{12}z_2)(b_{11}w_1 + b_{12}w_2) \\ &= a_{11}b_{11}(z_1w_1) + a_{11}b_{12}(z_1w_2) + a_{12}b_{11}(z_2w_1) + a_{12}b_{12}(z_2w_2) . \end{aligned}$$

Similar expressions for the other components lead to the transformation

$$\boldsymbol{\mu} = \begin{bmatrix} a_{11}b_{11} & a_{11}b_{12} & a_{12}b_{11} & a_{12}b_{12} \\ a_{11}b_{21} & a_{11}b_{22} & a_{12}b_{21} & a_{12}b_{22} \\ a_{21}b_{11} & a_{21}b_{12} & a_{22}b_{11} & a_{22}b_{12} \\ a_{21}b_{12} & a_{21}b_{22} & a_{22}b_{21} & a_{22}b_{22} \end{bmatrix} \mathbf{v}$$

or

$$\boldsymbol{\mu} = (A \otimes B)\mathbf{v} ,$$

that is

$$A\mathbf{z} \otimes B\mathbf{w} = (A \otimes B)(\mathbf{z} \otimes \mathbf{w}) . \tag{2.4}$$

Example 2.1

Let E_{ij} be an elementary matrix of order (2×2) defined in section 1.2 (see 1.4). Find the matrix

$$U = \sum_{i=1}^{2} \sum_{j=1}^{2} E_{i,j} \otimes E_{j,i} .$$

Solution

$$U = E_{11} \otimes E_{11} + E_{1,2} \otimes E_{2,1} + E_{21} \otimes E_{12} + E_{2,2} \otimes E_{2,2}$$

$$= \begin{bmatrix} 1 & 0 \\ 0 & 0 \end{bmatrix} \otimes \begin{bmatrix} 1 & 0 \\ 0 & 0 \end{bmatrix} + \begin{bmatrix} 0 & 1 \\ 0 & 0 \end{bmatrix} \otimes \begin{bmatrix} 0 & 0 \\ 1 & 0 \end{bmatrix} + \begin{bmatrix} 0 & 0 \\ 1 & 0 \end{bmatrix} \otimes \begin{bmatrix} 0 & 1 \\ 0 & 0 \end{bmatrix} + \begin{bmatrix} 0 & 0 \\ 0 & 1 \end{bmatrix} \otimes \begin{bmatrix} 0 & 0 \\ 0 & 1 \end{bmatrix}$$

so that

$$U = \begin{bmatrix} 1 & 0 & 0 & 0 \\ 0 & 0 & 1 & 0 \\ 0 & 1 & 0 & 0 \\ 0 & 0 & 0 & 1 \end{bmatrix} .$$

Note. U is seen to be a square matrix having columns which are unit vectors $\mathbf{e}_i (i = 1, 2, \ldots)$. It can be obtained from a unit matrix by a permutation of rows or columns. It is known as a **permutation matrix** (see also section 2.5).

2.3 SOME PROPERTIES AND RULES FOR KRONECKER PRODUCTS

We expect the Kronecker product to have the usual properties of a product.

I If α is a scalar, then

$$A \otimes (\alpha B) = \alpha(A \otimes B) . \tag{2.5}$$

Proof

The (i,j)th block of $A \otimes (\alpha B)$ is

$$[a_{ij}(\alpha B)]$$
$$= \alpha[a_{ij}B]$$
$$= \alpha[(i,j)\text{th block of } A \otimes B]$$

The result follows.

II The product is distributive with respect to addition, that is

$$\text{(a)} \quad (A + B) \otimes C = A \otimes C + B \otimes C \tag{2.6}$$

$$\text{(b)} \quad A \otimes (B + C) = A \otimes B + A \otimes C \tag{2.7}$$

Proof

We will only consider (a). The (i,j)th block of $(A + B) \otimes C$ is

$$(a_{ij} + b_{ij})C .$$

The (i,j)th block of $A \otimes C + B \otimes C$ is

$$a_{ij}C + b_{ij}C = (a_{ij} + b_{ij})C .$$

Since the two blocks are equal for every (i,j), the result follows.

III The product is associative

$$A \otimes (B \otimes C) = (A \otimes B) \otimes C \; . \tag{2.8}$$

IV There exists

$$\begin{aligned} &\text{a zero element } 0_{mn} = 0_m \otimes 0_n \\ &\text{a unit element } I_{mn} = I_m \otimes I_n \; . \end{aligned} \tag{2.9}$$

The unit matrices are all square, for example I_m in the unit matrix of order $(m \times m)$.

Other important properties of the Kronecker product follow.

V $(A \otimes B)' = A' \otimes B'$ (2.10)

Proof

The (i,j)th block of $(A \otimes B)'$ is

$$a_{ji}B' \; .$$

VI (The 'Mixed Product Rule').

$$(A \otimes B)(C \otimes D) = AC \otimes BD \tag{2.11}$$

provided the dimensions of the matrices are such that the various expressions exist.

Proof

The (i,j)th block of the left hand side is obtained by taking the product of the ith row block of $(A \otimes B)$ and the jth column block of $(C \otimes D)$, this is of the following form

$$[a_{i1}B \; a_{i2}B \; \ldots a_{in}B] \begin{bmatrix} c_{1j}D \\ c_{2j}D \\ \vdots \\ c_{nj}D \end{bmatrix}$$

$$= \sum_r a_{ir}c_{rj}BD \; .$$

The (i,j)th block of the right hand side is (by definition of the Kronecker product)

$$g_{ij}BD$$

where g_{ij} is the (i,j)th element of the matrix AC. But by the rule of matrix multiplications

$$g_{ij} = \sum_r a_{ir}c_{rj} \; .$$

Since the (i,j)th blocks are equal, the result follows.

VII Given $A(m \times m)$ and $B(n \times n)$ and subject to the existence of the various inverses,

$$(A \otimes B)^{-1} = A^{-1} \otimes B^{-1} \tag{2.12}$$

Proof

Use (2.11)

$$(A \otimes B)(A^{-1} \otimes B^{-1}) = AA^{-1} \otimes BB^{-1} = I_m \otimes I_n = I_{mn}$$

The result follows.

VIII (See (1.47))

$$\text{vec}(AYB) = (B' \otimes A)\,\text{vec}\,Y \tag{2.13}$$

Proof

We prove (2.13) for A, Y and B each of order $n \times n$. The result is true for $A(m \times n)$, $Y(n \times r)$, $B(r \times s)$. We use the solutions to Example 1.3(iii).

$$\begin{aligned}(AYB)_{\cdot k} &= \sum_j (b_{jk}A)Y_{\cdot j} \\ &= [b_{1k}A \; b_{2k}A \; \ldots \; b_{nk}A] \begin{bmatrix} Y_{\cdot 1} \\ Y_{\cdot 2} \\ \vdots \\ Y_{\cdot n} \end{bmatrix} \\ &= [B_{\cdot k}{}' \otimes A]\,\text{vec}\,Y \\ &= [(B')_{k\cdot}{}' \otimes A]\;\text{vec}\,Y\end{aligned}$$

since the transpose of the kth column of B is the kth row of B'; the results follows.

Example 2.2

Write the equation

$$\begin{bmatrix} a_{11} & a_{12} \\ a_{21} & a_{22} \end{bmatrix} \begin{bmatrix} x_1 & x_3 \\ x_2 & x_4 \end{bmatrix} = \begin{bmatrix} c_{11} & c_{12} \\ c_{21} & c_{22} \end{bmatrix}$$

in a matrix-vector form.

Solution

The equation can be written as $AXI = C$. Use (2.12), to find

$$\text{vec}(AXI) = (I \otimes A)\,\text{vec}\,X = \text{vec}\,C\,,$$

so that

$$\begin{bmatrix} a_{11} & a_{12} & 0 & 0 \\ a_{21} & a_{22} & 0 & 0 \\ 0 & 0 & a_{11} & a_{12} \\ 0 & 0 & a_{21} & a_{22} \end{bmatrix} \begin{bmatrix} x_1 \\ x_2 \\ x_3 \\ x_4 \end{bmatrix} = \begin{bmatrix} c_{11} \\ c_{21} \\ c_{12} \\ c_{22} \end{bmatrix}$$

Example 2.3

A and B are both of order $(n \times n)$, show that

(i) $\text{vec}\, AB = (I \otimes A)\,\text{vec}\, B$

(ii) $\text{vec}\, AB = (B' \otimes A)\,\text{vec}\, I$

(iii) $\text{vec}\, AB = \Sigma (B')_{\cdot k} \otimes A_{\cdot k}$

Solution

(i) (As in Example 2.2)
In (2.13) let $Y = B$ and $B = I$.

(ii) In (2.13) let $Y = I$.

(iii) In $\text{vec}\, AB = (B' \otimes A)\,\text{vec}\, I$

substitute (1.25), to obtain

$$\begin{aligned} \text{vec}\, AB &= \left[\sum_i (B')_{\cdot i} \mathbf{e}'_i \otimes \sum_j A_{\cdot j} \mathbf{e}'_j\right] \text{vec}\, I \\ &= \left[\sum_i \sum_j ((B')_{\cdot i} \otimes A_{\cdot j})(\mathbf{e}'_i \otimes \mathbf{e}'_j)\right] \text{vec}\, I \qquad \text{(by 2.11)} \end{aligned}$$

The product $\mathbf{e}'_i \otimes \mathbf{e}'_j$ is a one row matrix having a unit element in the $[(i-1)n + j]$th column and zeros elsewhere. Hence the product

$$[(B')_{\cdot i} \otimes A_{\cdot j}]\,[\mathbf{e}'_i \otimes \mathbf{e}'_j]$$

is a matrix having

$$(B')_{\cdot i} \otimes A_{\cdot j}$$

as its $[(i-1)n + j]$th column and zeros elsewhere. Since $\text{vec}\, I$ is a one column matrix having a unity in the 1st, $(n + 2)$nd, $(2n + 3)$rd ... n^2rd position and zeros elsewhere, the product of

$$[(B')_{\cdot i} \otimes A_{\cdot j}]\,[\mathbf{e}'_i \otimes \mathbf{e}'_j] \text{ and } \text{vec}\, I$$

is a one column matrix whose elements are all zeros unless i and j satisfy

$$(i-1)n + j = 1, \text{ or } n + 2, \text{ or } 2n + 3, \ldots, \text{ or } n^2$$

that is

$$i = j = 1 \quad \text{or} \quad i = j = 2 \quad \text{or} \quad i = j = 3 \quad \text{or} \quad \ldots, \quad i = j = n$$

in which case the one column matrix is

$$(B')_{\cdot i} \otimes A_{\cdot i} \quad (i = 1, 2, \ldots, n) \ .$$

The result now follows.

IX If $\{\lambda_i\}$ and $\{\mathbf{x}_i\}$ are the eigenvalues and the corresponding eigenvectors for A and $\{\mu_j\}$ and $\{\mathbf{y}_j\}$ are the eigenvalues and the corresponding eigenvectors for B, then

$$A \otimes B$$

has eigenvalues $\{\lambda_i\mu_j\}$ with corresponding eigenvectors $\{\mathbf{x}_i \otimes \mathbf{y}_j\}$.

Proof

By (2.11)

$$\begin{aligned}(A \otimes B)(\mathbf{x}_i \otimes \mathbf{y}_j) &= (A\mathbf{x}_i) \otimes (B\mathbf{y}_j) \\ &= (\lambda_i\mathbf{x}_i) \otimes (\mu_j\mathbf{y}_j) \\ &= \lambda_i\mu_j(\mathbf{x}_i \otimes \mathbf{y}_j) \qquad \text{(by 2.5)}\end{aligned}$$

The result follows.

X Given the two matrices A and B of order $n \times n$ and $m \times m$ respectively

$$|A \otimes B| = |A|^m |B|^n$$

where $|A|$ means the determinant of A.

Proof

Assume that $\lambda_1, \lambda_2, \ldots, \lambda_n$ and $\mu_1, \mu_2, \ldots, \mu_m$ are the eigenvalues of A and B respectively. The proof relies on the fact (see [18] p. 145) that the determinant of a matrix is equal to the product of its eigenvalues.

Hence (from Property IX above)

$$\begin{aligned}|A \otimes B| &= \prod_{i,j} \lambda_i\mu_j \\ &= \left(\lambda_1^m \prod_{j=1}^{n} \mu_j\right)\left(\lambda_2^m \prod_{j=1}^{n} \mu_j\right) \cdots \left(\lambda_n^m \prod_{j=1}^{n} \mu_j\right) \\ &= (\lambda_1\lambda_2 \ldots \lambda_n)^m (\mu_1\mu_2 \ldots \mu_m)^n \\ &= |A|^m |B|^n \ .\end{aligned}$$

Another important property of Kronecker products follows.

$$A \otimes B = U_1(B \otimes A)U_2 \tag{2.14}$$

where U_1 and U_2 are permutation matrices (see Example 2.1).

Proof

Let $AYB' = X$, then by (2.13)

$$(B \otimes A) \operatorname{vec} Y = \operatorname{vec} X . \qquad (1)$$

On taking transpose, we obtain

$$BY'A' = X'$$

So that by (2.13)

$$(A \otimes B) \operatorname{vec} Y' = \operatorname{vec} X' . \qquad (2)$$

From example 1.5, we know that there exist permutation matrices U_1 and U_2 such that

$$\operatorname{vec} X' = U_1 \operatorname{vec} X \quad \text{and} \quad \operatorname{vec} Y = U_2 \operatorname{vec} Y' .$$

Substituting for vec Y in (1) and multiplying both sides by U_1, we obtain

$$U_1(B \otimes A)U_2 \operatorname{vec} Y' = U_1 \operatorname{vec} X . \qquad (3)$$

Substituting for vec X' in (2), we obtain

$$(A \otimes B) \operatorname{vec} Y' = U_1 \operatorname{vec} X . \qquad (4)$$

The result follows from (3) and (4).

We will obtain an explicit formula for the permutation matrix U in section 2.5. Notice that U_1 and U_2 are independent of A and B except for the orders of the matrices.

XII If f is an analytic function, A is a matrix of order $(n \times n)$ and $f(A)$ exists, then

$$f(I_m \otimes A) = I_m \otimes f(A) \tag{2.15}$$

and

$$f(A \otimes I_m) = f(A) \otimes I_m \tag{2.16}$$

Proof

Since f is an analytic function it can be expressed as a power series such as

$$f(z) = a_0 + a_1 z + a_2 z^2 + \ldots$$

so that

$$f(A) = a_0 I_n + a_1 A + a_2 A^2 + \ldots = \sum_{k=0} a_k A^k ,$$

where $A^0 = I_n$.

By Cayley Hamilton's theorem (see [18]) the right hand side of the equation for $f(A)$ is the sum of at most $(n + 1)$ matrices.

We now have

$$f(I_m \otimes A) = \sum_{k=0} a_k (I_m \otimes A)$$

$$= \sum_{k=0} (I_m \otimes a_k A^k) \qquad \text{by (2.11)}$$

$$= \sum_{k=0} (I_m \otimes a_k A^k) \qquad \text{by (2.5)}$$

$$= I_m \otimes \sum_{k=0} a_k A^k \qquad \text{by (2.7)}$$

$$= I_m \otimes f(A) \ .$$

This proves (2.15); (2.16) is proved similarly.

We can write

$$f(A \otimes I_m) = \sum_{k=0} a_k (A \otimes I_m)^k$$

$$= \sum_{k=0} a_k (A^k \otimes I_m) \qquad \text{by (2.11)}$$

$$= \sum_{k=0} (a_k A^k \otimes I_m)$$

$$= \sum_{k=0} a_k A^k \quad \otimes I_m \qquad \text{by (2.6)}$$

$$= f(A) \otimes I_m$$

This proves (2.16).

An important application of the above property is for

$$f(z) = e^z \ .$$

(2.15) leads to the result

$$e^{I_m \otimes A} = I_m \otimes e^A \qquad (2.17)$$

and (2.16) leads to

$$e^{A \otimes I_m} = e^A \otimes I_m \qquad (2.18)$$

Example 2.4

Use a direct method to verify (2.17) and (2.18).

Solution

$$e^{I_m \otimes A} = I_m \otimes I_n + I_m \otimes A + \frac{1}{2!}(I_m \otimes A)^2 + \ldots$$

The right hand side is a block diagonal matrix, each of the m blocks is the sum

$$I_m + A + \frac{1}{2!}A^2 + \ldots = e^A \,.$$

The result (2.17) follows.

$$\begin{aligned} e^{A \otimes I_m} &= (I_n \otimes I_m) + (A \otimes I_m) + \frac{1}{2!}(I_m \otimes A)^2 + \ldots \\ &= (I_n \otimes I_m) + (A \otimes I_m) + \frac{1}{2!}(A^2 \otimes I_m) + \ldots \\ &= (I_n + A + \frac{1}{2!}A^2 + \ldots) \otimes I_m \\ &= e^A \otimes I_m \end{aligned}$$

XIII $\operatorname{tr}(A \otimes B) = \operatorname{tr} A \operatorname{tr} B$

Proof

Assume that A is of order $(n \times n)$

$$\begin{aligned} \operatorname{tr}(A \otimes B) &= \operatorname{tr}(a_{11}B) + \operatorname{tr}(a_{22}B) + \ldots + \operatorname{tr}(a_{nn}B) \\ &= a_{11}\operatorname{tr} B + a_{22}\operatorname{tr} B + \ldots + a_{nn}\operatorname{tr} B \\ &= (a_{11} + a_{22} + \ldots + a_{nn})\operatorname{tr} B \\ &= \operatorname{tr} A \operatorname{tr} B \,. \end{aligned}$$

2.4 DEFINITION OF THE KRONECKER SUM

Given a matrix $A(n \times n)$ and a matrix $B(m \times m)$, their **Kronecker Sum** denoted by $A \oplus B$ is defined as the expression

$$A \oplus B = A \otimes I_m + I_n \otimes B \tag{2.19}$$

We have seen (Property IX) that if $\{\lambda_i\}$ and $\{\mu_j\}$ are the eigenvalues of A and B respectively, then $\{\lambda_i \mu_j\}$ are the eigenvalues of the product $A \otimes B$. We now show the equivalent and fundamental property for $A \oplus B$.

XIV If $\{\lambda_i\}$ and $\{\mu_j\}$ are the eigenvalues of A and B respectively, then $\{\lambda_i + \mu_j\}$ are the eigenvalues of $A \oplus B$.

Proof

Let $\mathbf{x}$ and $\mathbf{y}$ be the eigenvectors corresponding to the eigenvalues λ and μ of A and B respectively, then

$$\begin{aligned} (A \oplus B)(\mathbf{x} \otimes \mathbf{y}) &= (A \otimes I)(\mathbf{x} \otimes \mathbf{y}) + (I \otimes B)(\mathbf{x} \otimes \mathbf{y}) && \text{by (2.19)} \\ &= (A\mathbf{x} \otimes \mathbf{y}) + (\mathbf{x} \otimes B\mathbf{y}) && \text{by (2.11)} \\ &= \lambda(\mathbf{x} \otimes \mathbf{y}) + \mu(\mathbf{x} \otimes \mathbf{y}) \\ &= (\lambda + \mu)(\mathbf{x} \otimes \mathbf{y}) \end{aligned}$$

The result follows.

Example 2.5

Verify the Property XIV for

$$A = \begin{bmatrix} 1 & -1 \\ 0 & 2 \end{bmatrix} \quad \text{and} \quad B = \begin{bmatrix} 1 & 0 \\ 2 & -1 \end{bmatrix}$$

Solution

For the matrix A;

$$\lambda_1 = 1 \quad \text{and} \quad \mathbf{x}_1 = \begin{bmatrix} 1 \\ 0 \end{bmatrix}$$

$$\lambda_2 = 2 \quad \text{and} \quad \mathbf{x}_2 = \begin{bmatrix} 1 \\ -1 \end{bmatrix} .$$

For the matrix B;

$$\mu_1 = 1 \quad \text{and} \quad \mathbf{y}_1 = \begin{bmatrix} 1 \\ 1 \end{bmatrix}$$

$$\mu_2 = -1 \quad \text{and} \quad \mathbf{y}_2 = \begin{bmatrix} 0 \\ 1 \end{bmatrix} .$$

We find

$$C = A \oplus B = \begin{bmatrix} 2 & 0 & -1 & 0 \\ 2 & 0 & 0 & -1 \\ 0 & 0 & 3 & 0 \\ 0 & 0 & 2 & 1 \end{bmatrix}$$

and $|pI - C| = p(p-1)(p-2)(p-3)$, so that the eigenvalues of $A \oplus B$ are

$$p = 0 = \lambda_1 + \mu_2 \quad \text{and} \quad \mathbf{x}_1 \otimes \mathbf{y}_2 = [0\ 1\ 0\ 0]'$$

$$p = 1 = \lambda_2 + \mu_2 \quad \text{and} \quad \mathbf{x}_2 \otimes \mathbf{y}_2 = [0\ 1\ 0\ -1]'$$

$$p = 2 = \lambda_1 + \mu_1 \quad \text{and} \quad \mathbf{x}_1 \otimes \mathbf{y}_1 = [1\ 1\ 0\ 0]'$$

and

$$p = 3 = \lambda_2 + \mu_1 \quad \text{and} \quad \mathbf{x}_2 \otimes \mathbf{y}_1 = [1\ 1\ -1\ -1]' .$$

The Kronecker sum frequently turns up when we are considering equations of the form;

$$AX + XB = C \tag{2.20}$$

where $A(n \times n)$, $B(m \times m)$ and $X(n \times m)$.

Use (2.13) and solution to Example 2.3 to write the above in the form

$$(I_m \otimes A + B' \otimes I_n) \operatorname{vec} X = \operatorname{vec} C \tag{2.21}$$

or

$$(B' \oplus A) \operatorname{vec} X = \operatorname{vec} C .$$

It is interesting to note the generality of the Kronecker sum. For example,

$$\exp(A + B) = \exp A \exp B$$

if and only if A and B commute (see [18] p. 227)
whereas $\exp(A \oplus B) = \exp(A \otimes I)\exp(I \otimes B)$
even if A and B do not commute!

Example 2.6
Show that

$$\exp(A \oplus B) = \exp A \otimes \exp B$$

where $A(n \times n), B(m \times m)$.

Solution
By (2.11)

$$(A \otimes I_m)(I_n \otimes B) = A \otimes B$$

and

$$(I_n \otimes B)(A \otimes I_m) = A \otimes B$$

hence $(A \otimes I_m)$ and $(I_n \otimes B)$ commute so that

$$\begin{aligned}\exp(A \oplus B) &= \exp(A \otimes I_m + I_n \otimes B)\\ &= \exp(A \otimes I_m)\exp(I_n \otimes B)\\ &= (\exp A \otimes I_m)(I_n \otimes \exp B) && \text{(by 2.15 and 2.16)}\\ &= \exp A \otimes \exp B && \text{(by 2.11)}\end{aligned}$$

2.5 THE PERMUTATION MATRIX ASSOCIATING vec X AND vec X'

If $X = [x_{ij}]$ is a matrix of order $(m \times n)$ we can write (see (1.20))

$$X = \sum_i \sum_j x_{ij} E_{ij}$$

where E_{ij} is an elementary matrix of order $(m \times n)$. It follows that

$$X' = \sum\sum x_{ij} E'_{ij}\,.$$

so that

$$\text{vec } X' = \sum\sum x_{ij} \text{ vec } E'_{ij}\,. \tag{2.22}$$

We can write (2.22) in a form of matrix multiplication as

$$\text{vec } X' = [\text{vec } E'_{11} \vdots \text{vec } E'_{21} \vdots \ldots \text{vec } E'_{m1} \vdots \text{vec } E'_{12} \vdots \ldots \text{vec } E'_{mn}] \begin{bmatrix} x_{11} \\ x_{21} \\ \vdots \\ x_{m1} \\ x_{12} \\ \vdots \\ x_{mn} \end{bmatrix}$$

that is

$$\text{vec}\, X' = [\text{vec}\, E'_{11} \vdots \text{vec}\, E'_{21} \vdots \ldots \text{vec}\, E'_{m1} \vdots \text{vec}\, E'_{12} \vdots \ldots \text{vec}\, E'_{mn}]\ \text{vec}\, X.$$

So the permutation matrix associating vec X and vec X' is

$$U = [\text{vec}\, E'_{11} \vdots \text{vec}\, E'_{21} \vdots \ldots \text{vec}\, E'_{mn}]\ . \tag{2.23}$$

Example 2.7

Given

$$X = \begin{bmatrix} x_{11} & x_{12} & x_{13} \\ x_{21} & x_{22} & x_{23} \end{bmatrix}, \quad \text{determine the matrix } U$$

such that

$$\text{vec}\, X' = U\, \text{vec}\, X.$$

Solution

$$E'_{11} = \begin{bmatrix} 1 & 0 \\ 0 & 0 \\ 0 & 0 \end{bmatrix},\ E'_{21} = \begin{bmatrix} 0 & 1 \\ 0 & 0 \\ 0 & 0 \end{bmatrix},\ E'_{12} = \begin{bmatrix} 0 & 0 \\ 1 & 0 \\ 0 & 0 \end{bmatrix},\ E'_{22} = \begin{bmatrix} 0 & 0 \\ 0 & 1 \\ 0 & 0 \end{bmatrix}$$

$$E'_{13} = \begin{bmatrix} 0 & 0 \\ 0 & 0 \\ 1 & 0 \end{bmatrix},\ \text{and}\ E'_{23} = \begin{bmatrix} 0 & 0 \\ 0 & 0 \\ 0 & 1 \end{bmatrix}.$$

Hence by (2.23)

$$U = \begin{bmatrix} 1 & 0 & 0 & 0 & 0 & 0 \\ 0 & 0 & 1 & 0 & 0 & 0 \\ 0 & 0 & 0 & 0 & 1 & 0 \\ 0 & 1 & 0 & 0 & 0 & 0 \\ 0 & 0 & 0 & 1 & 0 & 0 \\ 0 & 0 & 0 & 0 & 0 & 1 \end{bmatrix}.$$

We now obtain the permutation matrix U in a useful form as a Kronecker product of elementry matrices.

As it is necessary to be precise about the suffixes of the elementary matrices, we will use the notation explained at the end of Chapter 1.

As above, we write

$$X' = \sum_{r=1}^{m} \sum_{s=1}^{n} x_{rs} E_{sr}\ (n \times m)\ .$$

By (1.31) we can write

$$X' = \sum_{r,s} E_{sr}\ (n \times m)\ X E_{sr}\ (n \times m)\ .$$

Hence,

$$\text{vec}\, X' = \text{vec} \sum_{r,s} E_{sr}\,(n \times m)\, X E_{sr}\,(n \times m)$$

$$= \sum_{r,s} [E_{rs}\,(m \times n) \otimes E_{sr}\,(n \times m)]\; \text{vec}\, X\;. \qquad \text{by (2.13)}$$

It follows that

$$U = \sum_{r,s} E_{rs}\,(m \times n) \otimes E_{sr}\,(n \times m) \qquad (2.24)$$

or in our less rigorous notation

$$U = \sum_{r,s} E_{rs} \otimes E'_{rs}\,. \qquad (2.25)$$

Notice that U is a matrix of order $(mn \times mn)$.

At first sight it may appear that the evaluation of the permutation matrices U_1 and U_2 in (2.14) using (2.24) is a major task. In fact this is one of the examples where the practice is much easier than the theory.

We can readily determine the form of a permutation matrix – as in Example 2.7. So the only real problem is to determine the orders of the two matrices.

Since the matrices forming the product (2.14) must be conformable, the orders of the matrices U_1 and U_2 are determined respectively by the number of rows and the number of columns of $(A \otimes B)$.

Example 2.8

Let $A = [a_{ij}]$ be a matrix of order (2×3), and $B = [b_{ij}]$ be a matrix of order (2×2).

Determine the permutation matrices U_1 and U_2 such that

$$A \otimes B = U_1\;(B \otimes A)\, U_2$$

Solution

$(A \otimes B)$ is of the order (4×6)

From the above discussion we conclude that U_1 is of order (4×4) and U_2 is of order (6×6).

$$U_1 = \begin{bmatrix} 1 & 0 & 0 & 0 \\ 0 & 0 & 1 & 0 \\ 0 & 1 & 0 & 0 \\ 0 & 0 & 0 & 1 \end{bmatrix} \quad \text{and} \quad U_2 = \begin{bmatrix} 1 & 0 & 0 & 0 & 0 & 0 \\ 0 & 0 & 1 & 0 & 0 & 0 \\ 0 & 0 & 0 & 0 & 1 & 0 \\ 0 & 1 & 0 & 0 & 0 & 0 \\ 0 & 0 & 0 & 1 & 0 & 0 \\ 0 & 0 & 0 & 0 & 0 & 1 \end{bmatrix}.$$

Another related matrix which will be used (in Chapter 6) is

$$\bar{U} = \sum_{r,s} E_{rs} \otimes E_{rs} \,. \tag{2.26}$$

When the matrix X is or order $(m \times n)$, $\bar{U}$ is or order $(m^2 \times n^2)$.

Problems for Chapter 2

(1) Given

$$U = \sum_{r,s} E_{rs}\,(m \times n) \otimes E_{sr}\,(n \times m) \,.$$

Show that

$$U^{-1} = U' = \sum_{r,s} E_{sr}\,(n \times m) \otimes E_{rs}\,(m \times n)$$

(2) $A = [a_{ij}]$, $B = [b_{ij}]$ and $Y = [y_{ij}]$ are matrices all of order (2×2), use a direct method to evaluate

(a) (i) AYB

(ii) $B' \otimes A$.

(b) Verify (2.13) that

$$\text{vec}\, AYB = (B' \otimes A)\, \text{vec}\, Y \,.$$

(3) Given

$$A = \begin{bmatrix} 2 & 1 \\ 0 & 1 \end{bmatrix} \quad \text{and} \quad B = \begin{bmatrix} -1 & 1 \\ 2 & 0 \end{bmatrix}$$

(a) Calculate

$$A \otimes B \quad \text{and} \quad B \otimes A \,.$$

(b) Find matrices U_1 and U_2 such that

$$A \otimes B = U_1 (B \otimes A) U_2 \,.$$

(4) Given

$$A = \begin{bmatrix} 3 & 4 \\ -2 & -3 \end{bmatrix}, \quad \text{calculate}$$

(a) $\exp(A)$

(b) $\exp(A \otimes I)$.

Verify (2.16), that is

$$\exp(A) \otimes I = \exp(A \otimes I).$$

(5) Given

$$A = \begin{bmatrix} 2 & 1 \\ -1 & -1 \end{bmatrix} \quad \text{and} \quad B = \begin{bmatrix} 1 & 2 \\ 3 & 4 \end{bmatrix}, \quad \text{calculate}$$

(a) $A^{-1} \otimes B^{-1}$ and

(b) $(A \otimes B)^{-1}$.

Hence verify (2.12), that is

$$(A \otimes B)^{-1} = A^{-1} \otimes B^{-1} .$$

(6) Given

$$A = \begin{bmatrix} 3 & -1 \\ 4 & -2 \end{bmatrix} \quad \text{and} \quad B = \begin{bmatrix} 2 & 1 \\ 2 & 3 \end{bmatrix}, \quad \text{find}$$

(a) The eigenvalues and eigenvectors of A and B.

(b) The eigenvalues and eigenvectors of $A \otimes B$.

(c) Verify Property IX of Kronecker Products.

(7) A, B, C and D are matrices such that

A is similar to C, and

B is similar to D.

Show that $A \otimes B$ is similar to $C \otimes D$.

CHAPTER 3

Some Applications of the Kronecker Product

3.1 INTRODUCTION

There are numerous applications of the Kronecker product in various fields including statistics, economics, optimisation and control. It is not our intention to discuss applications in all these fields, just a selected number to give an idea of the problems tackled in some of the literature mentioned in the Bibliography. There is no doubt that the interested reader will find there various other applications hopefully in his own field of interest.

A number of the applications involve the derivative of a matrix – it is a well known concept (for example see [18] p. 229) which we now briefly review.

3.2 THE DERIVATIVE OF A MATRIX

Given the matrix

$$A(t) = [a_{ij}(t)]$$

the derivative of the matrix, with respect to a scalar variable t, denoted by $(\mathrm{d}/\mathrm{d}t)A(t)$ or just $\mathrm{d}A/\mathrm{d}t$ or $\dot{A}(t)$ is defined as the matrix

$$\frac{\mathrm{d}}{\mathrm{d}t}A(t) = \left[\frac{\mathrm{d}}{\mathrm{d}t}a_{ij}(t)\right] . \tag{3.1}$$

Similarly, the integral of the matrix is defined as

$$\int A(t)\mathrm{d}t = \left[\int a_{ij}(t)\mathrm{d}t\right] . \tag{3.2}$$

For example, given

$$A = \begin{bmatrix} 2t^2 & 4 \\ \sin t & 2+t^2 \end{bmatrix}$$

then

$$\frac{\mathrm{d}}{\mathrm{d}t}A = \begin{bmatrix} 4t & 0 \\ \cos t & 2t \end{bmatrix} \quad \text{and} \quad \int A\,\mathrm{d}t = \begin{bmatrix} \frac{2}{3}t^3 & 4t \\ -\cos t & 2t + t^3/3 \end{bmatrix} + C$$

where C is a constant matrix.

One important property follows immediately. Given conformable matrices $A(t)$ and $B(t)$, then

$$\frac{\mathrm{d}}{\mathrm{d}t}[AB] = \frac{\mathrm{d}A}{\mathrm{d}t}B + A\frac{\mathrm{d}B}{\mathrm{d}t} \quad . \tag{3.3}$$

Example 3.1

Given

$$C = A \otimes B$$

(each matrix is assumed to be a function of t) show that

$$\frac{\mathrm{d}C}{\mathrm{d}t} = \frac{\mathrm{d}A}{\mathrm{d}t} \otimes B + A \otimes \frac{\mathrm{d}B}{\mathrm{d}t} \tag{3.4}$$

Solution

On differentiating the (i,j)th block of $A \otimes B$, we obtain

$$\frac{\mathrm{d}}{\mathrm{d}t}(a_{ij}B) = \frac{\mathrm{d}a_{ij}}{\mathrm{d}t}B + a_{ij}\frac{\mathrm{d}B}{\mathrm{d}t}$$

which is the (i,j)th partition of

$$\frac{\mathrm{d}A}{\mathrm{d}t} \otimes B + A \otimes \frac{\mathrm{d}B}{\mathrm{d}t} \, ,$$

the result follows.

3.3 PROBLEM 1

Determine the condition for the equation

$$AX + XB = C$$

to have a unique solution.

Solution

We have already considered this equation and wrote it (2.21) as

$$(B' \oplus A) \operatorname{vec} X = \operatorname{vec} C$$

or

$$G\mathbf{x} = \mathbf{c} \tag{3.5}$$

where $G = B' \oplus A$ and $\mathbf{c} = \operatorname{vec} C$.

Equation (3.5) has a unique solution iff G is nonsingular, that is iff the eigenvalues of G are all nonzero. Since, by Property XIV (see section 2.4), the eigenvalues of G are $\{\lambda_i + \mu_j\}$ (note that the eigenvalues of the matrix B' are the same as the eigenvalues of B). Equation (3.5) has a unique solution iff

$$\lambda_i + \mu_j \neq 0 \quad \text{(all } i \text{ and } j\text{)} .$$

We have thus proved that $AX + BX = C$ has a unique solution iff A and $(-B)$ have no eigenvalue in common.

If on the other hand, A and $(-B)$ have common eigenvalues then the existence of solutions depends on the rank of the augmented matrix

$$[G \vdots \mathbf{c}] \ .$$

If the rank of $[G \vdots \mathbf{c}]$ is equal to the rank of G, then solutions do exist, otherwise the set of equations

$$AX + XB = C$$

is not consistent.

Example 3.2

Obtain the solution to

$$AX + XB = C$$

where

(i) $$A = \begin{bmatrix} 1 & -1 \\ 0 & 2 \end{bmatrix}, \quad B = \begin{bmatrix} -3 & 4 \\ 1 & 0 \end{bmatrix} \quad \text{and} \quad C = \begin{bmatrix} 1 & 3 \\ -2 & 2 \end{bmatrix} .$$

(ii) $$A = \begin{bmatrix} 1 & -1 \\ 0 & 2 \end{bmatrix}, \quad B = \begin{bmatrix} -3 & 4 \\ 0 & -1 \end{bmatrix} \quad \text{and} \quad C = \begin{bmatrix} 0 & 5 \\ 2 & -9 \end{bmatrix} .$$

Solution

Writing the equation in the form of (3.5) we obtain,

(i) $$\begin{bmatrix} -2 & -1 & 1 & 0 \\ 0 & -1 & 0 & 1 \\ 4 & 0 & 1 & -1 \\ 0 & 4 & 0 & 2 \end{bmatrix} \begin{bmatrix} x_1 \\ x_2 \\ x_3 \\ x_4 \end{bmatrix} = \begin{bmatrix} 1 \\ -2 \\ 3 \\ 2 \end{bmatrix}$$

where for convenience we have denoted

$$X = \begin{bmatrix} x_1 & x_3 \\ x_2 & x_4 \end{bmatrix} .$$

On solving we obtain the unique solution

$$X = \begin{bmatrix} 0 & 2 \\ 1 & -1 \end{bmatrix} .$$

(ii) In case (ii) A and $(-B)$ have one eigenvalue ($\lambda = 1$) in common. Equation (3.5) becomes

$$\begin{bmatrix} -2 & -1 & 0 & 0 \\ 0 & -1 & 0 & 0 \\ 4 & 0 & 0 & -1 \\ 0 & 4 & 0 & 1 \end{bmatrix} \begin{bmatrix} x_1 \\ x_2 \\ x_3 \\ x_4 \end{bmatrix} = \begin{bmatrix} 0 \\ 2 \\ 5 \\ -9 \end{bmatrix}$$

and rank G = rank $[G \vdots \mathbf{c}]$.

G is seen to be singular, but

$$\text{rank } G = \text{rank } [G \vdots \mathbf{c}] = 3$$

hence at least one solution exists. In fact two linearly independent solutions are

$$X_1 = \begin{bmatrix} 1 & 0 \\ -2 & -1 \end{bmatrix} \quad \text{and} \quad X_2 = \begin{bmatrix} 1 & 1 \\ -2 & -1 \end{bmatrix}$$

any other solution is a linear combination of X_1 and X_2.

3.4 PROBLEM 2

Determine the condition for the equation

$$AX - XA = \mu X \tag{3.6}$$

to have a nontrivial solution.

Solution

We can write (3.6) as

$$H\mathbf{x} = \mu\mathbf{x} \tag{3.7}$$

where $H = I \otimes A - A' \otimes I$ and

$$\mathbf{x} = \text{vec } X .$$

(3.7) has a nontrivial solution for $\mathbf{x}$ iff

$$|\mu I - H| = 0$$

that is iff μ is an eigenvalue of H. But by a simple generalisation of Property XIV, section 2.4, the eigenvalues of H are $\{(\lambda_i - \lambda_j)\}$ where $\{\lambda_i\}$ are the eigenvalues of A. Hence (3.6) has a nontrivial solution iff

$$\mu = \lambda_i - \lambda_j \ .$$

Example 3.3

Determine the solutions to (3.6) when

$$A = \begin{bmatrix} 1 & 0 \\ 2 & 3 \end{bmatrix} \quad \text{and} \quad \mu = -2 \ .$$

Solution

$\mu = -2$ is an eigenvalue of H, hence we expect a nontrivial solution. Equation (3.7) becomes

$$\begin{bmatrix} 0 & 0 & -2 & 0 \\ 2 & 2 & 0 & -2 \\ 0 & 0 & -2 & 0 \\ 0 & 0 & 2 & 0 \end{bmatrix} \begin{bmatrix} x_1 \\ x_2 \\ x_3 \\ x_4 \end{bmatrix} = -2 \begin{bmatrix} x_1 \\ x_2 \\ x_3 \\ x_4 \end{bmatrix}$$

On solving, we obtain

$$X = \begin{bmatrix} 1 & 1 \\ -1 & -1 \end{bmatrix} .$$

3.5 PROBLEM 3

Use the fact (see [18] p. 230) that the solution to

$$\dot{\mathbf{x}} = A\mathbf{x} \ , \quad \mathbf{x}(0) = \mathbf{c} \tag{3.8}$$

is

$$\mathbf{x} = \exp(At)\,\mathbf{c} \tag{3.9}$$

to solve the equation

$$\dot{X} = AX + XB \ , \quad X(0) = C \tag{3.10}$$

where $A(n \times n)$, $B(m \times m)$ and $X(n \times m)$.

Solution

Using the vec operator on (3.10) we obtain

$$\dot{\mathbf{x}} = G\mathbf{x} \ , \quad \mathbf{x}(0) = \mathbf{c} \tag{3.11}$$

where

$$\mathbf{x} = \text{vec}\, X \ , \quad \mathbf{c} = \text{vec}\, C$$

and

$$G = I_m \otimes A + B' \otimes I_n \ .$$

By (3.9) the solution to (3.11) is

$$\begin{aligned}\operatorname{vec} X &= \exp\{(I_m \otimes A)t + (B' \otimes I_n)t\} \operatorname{vec} C \\ &= [\exp(I_m \otimes A)t][\exp(B' \otimes I_n)t] \operatorname{vec} C \quad \text{(see Example 2.6)} \\ &= [I_m \otimes \exp(At)][\exp(B't) \otimes I_n] \operatorname{vec} C \quad \text{by (2.17) and (2.18).}\end{aligned}$$

We now make use of the result

$$\operatorname{vec} AB = (B' \otimes I) \operatorname{vec} A$$

(in (2.13) put $A = I$ and $Y \dot{=} A$) in conjunction with the fact that

$$[\exp(B't)]' = \exp(Bt),$$

to obtain

$$(\exp(B't) \otimes I_n) \operatorname{vec} C = \operatorname{vec}[C \exp(Bt)] .$$

Using the result of Example 2.3(1), we finally obtain

$$\operatorname{vec} X = \operatorname{vec}[\exp(At)\, C \exp(Bt) \tag{3.12}$$

So that $X = \exp(At)\, C \exp(Bt)$.

Example 3.4

Obtain the solution to (3.10) when

$$A = \begin{bmatrix} 1 & -1 \\ 0 & 2 \end{bmatrix}, \quad B = \begin{bmatrix} 1 & 0 \\ 0 & -1 \end{bmatrix} \quad \text{and} \quad C = \begin{bmatrix} -2 & 0 \\ 1 & 1 \end{bmatrix}.$$

Solution

(See [18] p. 227)

$$\exp(At) = \begin{bmatrix} e^t & e^t - e^{2t} \\ 0 & e^{2t} \end{bmatrix}, \quad \exp(Bt) = \begin{bmatrix} e^t & 0 \\ 0 & e^{-t} \end{bmatrix}$$

hence

$$X = \begin{bmatrix} -e^{2t} - e^{3t} & 1 - e^t \\ e^{3t} & e^t \end{bmatrix}.$$

3.6 PROBLEM 4

We consider a problem similar to the previous one but in a different context.

An important concept in Control Theory is the **transition** matrix.

Very briefly, associated with the equations

$$\dot{X} = A(t)X \quad \text{or} \quad \dot{\mathbf{x}} = A(t)\mathbf{x}$$

is the transition matrix $\Phi_1(t, \tau)$ having the following two properties

$$\dot{\Phi}_1(t, \tau) = A(t)\Phi_1(t, \tau) \tag{3.13}$$

and

$$\Phi_1(t, t) = I$$

[For simplicity of notation we shall write Φ for $\Phi(t,\tau)$.] If A is a constant matrix, it is easily shown that

$$\Phi = \exp(At) \ .$$

Similarly, with the equation

$$\dot{X} = XB \quad \text{so that} \quad \dot{X}' = B'X'$$

we associate the transition matrix Φ_2 such that

$$\dot{\Phi}_2 = B'\Phi_2 \ . \tag{3.14}$$

The problem is to find the transition matrix associated with the equation

$$\dot{X} = AX + XB \tag{3.15}$$

given the transition matrices Φ_1 and Φ_2 defined above.

Solution

We can write (3.15) as

$$\dot{\mathbf{x}} = G\mathbf{x}$$

where $\mathbf{x}$ and G were defined in the previous problem.

We define a matrix ψ as

$$\psi(t,\tau) = \Phi_2(t,\tau) \otimes \Phi_1(t,\tau) \ . \tag{3.16}$$

We obtain by (3.4)

$$\begin{aligned} \dot{\psi} &= \dot{\Phi}_2 \otimes \Phi_1 + \Phi_2 \otimes \dot{\Phi}_1 \\ &= (B'\Phi_2) \otimes \Phi_1 + \Phi_2 \otimes (A\Phi_1) && \text{by (3.13) and (3.14)} \\ &= (B'\Phi_2) \otimes (I\Phi_1) + (I\Phi_2) \otimes (A\Phi_1) \\ &= [B' \otimes I + I \otimes A][\Phi_2 \otimes \Phi_1] \ . && \text{by (2.11)} \end{aligned}$$

Hence

$$\dot{\psi} = G\psi \ . \tag{3.17}$$

Also

$$\begin{aligned} \psi(t,t) &= \Phi_2(t,t) \otimes \Phi_1(t,t) \\ &= I \otimes I \\ &= I \ . \end{aligned} \tag{3.18}$$

The two equations (3.17) and (3.18) prove that ψ is the transition matrix for (3.15)

Example 3.5

Find the transition matrix for the equation

$$\dot{X} = \begin{bmatrix} 1 & -1 \\ 0 & 2 \end{bmatrix} X + X \begin{bmatrix} 1 & 0 \\ 0 & -1 \end{bmatrix} .$$

Solution

In this case both A and B are constant matrices. From Example 3.4.

$$\Phi_1 = \exp(At) = \begin{bmatrix} e^t & e^t - e^{2t} \\ 0 & e^{2t} \end{bmatrix}$$

$$\Phi_2 = \exp(Bt) = \begin{bmatrix} e^t & 0 \\ 0 & e^{-t} \end{bmatrix}$$

so that

$$\psi = \Phi_2 \otimes \Phi_1 = \begin{bmatrix} e^{2t} & e^{2t} - e^{3t} & 0 & 0 \\ 0 & e^{3t} & 0 & 0 \\ 0 & 0 & 1 & 1 - e^t \\ 0 & 0 & 0 & e^t \end{bmatrix}.$$

For this equation

$$G = \begin{bmatrix} 2 & -1 & 0 & 0 \\ 0 & 3 & 0 & 0 \\ 0 & 0 & 0 & -1 \\ 0 & 0 & 0 & 1 \end{bmatrix}$$

and it is easily verified that

$$\dot{\psi} = G\psi$$

and

$$\psi(0) = I.$$

3.7 PROBLEM 5

Solve the equation

$$AXB = C$$

where all matrices are of order $n \times n$.

Solution

Using (2.13) we can write (3.19) in the form

$$H\mathbf{x} = \mathbf{c} \tag{3.20}$$

where $H = B' \otimes A$, $\mathbf{x} = \text{vec}\, X$ and $\mathbf{c} = \text{vec}\, C$.

The criteria for the existence and the uniqueness of a solution to (3.20) are well known (see for example [18]).

The above method of solving the problem is easily generalised to the linear equation of the form

$$A_1XB_1 + A_2XB_2 + \ldots + A_rXB_r = C \tag{3.21}$$

Equation (3.21) can be written as for example (3.20) where this time

$$H = B_1' \otimes A_1 + B_2' \otimes A_2 + \ldots + B_r' \otimes A_r \ .$$

Example 3.6

Find the matrix X, given

$$A_1 X B_1 + A_2 X B_2 = C$$

where

$$A_1 = \begin{bmatrix} 2 & 2 \\ 2 & -1 \end{bmatrix}, \quad B_1 = \begin{bmatrix} 1 & 0 \\ -1 & 1 \end{bmatrix}, \quad A_2 = \begin{bmatrix} 0 & 1 \\ -2 & -1 \end{bmatrix},$$

$$B_2 = \begin{bmatrix} 0 & 2 \\ -1 & 3 \end{bmatrix}, \quad \text{and} \quad C = \begin{bmatrix} 4 & -6 \\ 0 & 8 \end{bmatrix}.$$

Solution

For this example it is found that

$$H = B_1' \otimes A_1 + B_2' \otimes A_2 = \begin{bmatrix} 2 & 2 & -2 & -3 \\ 2 & -1 & 0 & 2 \\ 0 & 2 & 2 & 5 \\ -4 & -2 & -4 & -4 \end{bmatrix}$$

and $\mathbf{c}' = [4 \ \ 0 \ \ -6 \ \ 8]$.

It follows that

$$\mathbf{x} = H^{-1}\mathbf{c} = \begin{bmatrix} 1 \\ -1 \\ -2 \\ 0 \end{bmatrix}$$

so that

$$X = \begin{bmatrix} 1 & -2 \\ -1 & 0 \end{bmatrix}$$

3.8 PROBLEM 6

This problem is to determine a constant output feedback matrix K so that the closed loop matrix of a system has preassigned eigenvalues.

A multivariable system is defined by the equations

$$\dot{\mathbf{x}} = A\mathbf{x} + B\mathbf{u}$$
$$\mathbf{y} = C\mathbf{x} \qquad (3.22)$$

where $A(n \times n)$, $B(n \times m)$ and $C(r \times n)$ are constant matrices. $\mathbf{u}$, $\mathbf{x}$ and $\mathbf{y}$ are column vectors of order m, n and r respectively.

We are concerned with a system having an output feedback law of the form

$$\mathbf{u} = K\mathbf{y} \tag{3.23}$$

where $K(m \times r)$ is the constant control matrix to be determined.

On substituting (3.23) into (3.22), we obtain the equations of the closed loop system

$$\begin{aligned} \dot{\mathbf{x}} &= (A+BKC)\mathbf{x} \\ \mathbf{y} &= C\mathbf{x} \ . \end{aligned} \tag{3.24}$$

The problem can now be restated as follows:

Given the matrices A, B, and C, determine a matrix K such that

$$\begin{aligned} |\lambda I - A - BKC| &= a_0 + a_1\lambda + \ldots + a_{n-1}\lambda^{n-1} + \lambda^n \text{ (say)} \\ &= 0 \text{ for preassigned values } \lambda = \lambda_1, \lambda_2, \ldots, \lambda_n \ . \end{aligned} \tag{3.25}$$

Solution

Various solutions exist to this problem. We are interested in the application of the Kronecker product and will follow a method suggested in [24].

We consider a matrix $H(n \times n)$ whose eigenvalues are the desired values $\lambda_1, \lambda_2, \ldots, \lambda_n$, that is

$$|\lambda I - H| = 0 \quad \text{for} \quad \lambda = \lambda_1, \lambda_2, \ldots, \lambda_n \tag{3.26}$$

and

$$|\lambda I - H| = a_0 + a_1\lambda + \ldots + a_{n-1}\lambda^{n-1} + \lambda^n \ . \tag{3.27}$$

Let

$$A + BKC = H$$

so that

$$BKC = H - A = Q \text{ (say)} \ . \tag{3.28}$$

Using (2.13) we can write (3.28) as

$$(C' \otimes B) \operatorname{vec} K = \operatorname{vec} Q \tag{3.29}$$

or more simply as

$$P\mathbf{k} = \mathbf{q} \tag{3.30}$$

where $P = C' \otimes B$, $\mathbf{k} = \operatorname{vec} K$ and $\mathbf{q} = \operatorname{vec} Q$.

Notice that P is of order $(n^2 \times mr)$ and $\mathbf{k}$ and $\mathbf{q}$ are column vectors of order mr and n^2 respectively.

The system of equations (3.30) is overdetermined unless of course $m = n = r$, in which case can be solved in the usual manner – assuming a solution does exist!

In general, to solve the system for $\mathbf{k}$ we must consider the subsystem of linearly independent equations, the remaining equations being linearly dependent on this subsystem. In other words we determine a nonsingular matrix $T(n^2 \times n^2)$ such that

$$TP = \begin{bmatrix} P_1 \\ \hdashline P_2 \end{bmatrix} \tag{3.31}$$

where P_1 is the matrix of the coefficients of the linearly independent equations of the system (3.30) and P_2 is a null matrix.

Premultiplying both sides of (3.30) by T and making use of (3.31), we obtain

$$TP\mathbf{k} = T\mathbf{q}$$

or

$$\begin{bmatrix} P_1 \\ \hdashline P_2 \end{bmatrix} \mathbf{k} = \begin{bmatrix} \mathbf{u} \\ \hdashline \mathbf{v} \end{bmatrix} \quad . \tag{3.32}$$

If the rank of P is mr, then P_1 is of order $(mr \times mr)$, P_2 is of order $([n^2 - mr] \times mr)$ and $\mathbf{u}$ and $\mathbf{v}$ are of order mr and $(n^2 - mr)$ respectively.

A sufficient condition for the existence of a solution to (3.32) or equivalently to (3.30) is that

$$\mathbf{v} = 0 \tag{3.33}$$

in (3.32).

If the condition (3.33) holds and rank $P_1 = mr$, then

$$\mathbf{k} = P_1^{-1}\mathbf{u} \quad . \tag{3.34}$$

The condition (3.33) depends on an appropriate choice of H. The underlying assumption being made is that a matrix H satisfying this condition does exist. This in turn depends on the system under consideration, for example whether it is controllable.

Some obvious choices for the form of matrix H are: (a) diagonal, (b) upper or lower triangular, (c) companion form or (d) certain combinations of the above forms.

Although forms (a) and (b) are well known, the companion form is less well documented.

Very briefly, the matrix

$$H = \begin{bmatrix} 0 & 1 & 0 & \dots & 0 \\ 0 & 0 & 1 & \dots & 0 \\ 0 & 0 & 0 & \dots & 1 \\ -a_0 & -a_1 & -a_2 & & -a_{n-1} \end{bmatrix}$$

is said to be in 'companion' form, it has the associated characteristic equation

$$|\lambda I - H| = a_0 + a_1\lambda + \dots + a_{n-1}\lambda^{n-1} + \lambda^n = 0 \tag{3.35}$$

Example 3.7

Determine the feedback matrix K so that the two input – two output system

$$\dot{\mathbf{x}} = \begin{bmatrix} 0 & 1 & 0 \\ 3 & 3 & 1 \\ 2 & -3 & 2 \end{bmatrix} \mathbf{x} + \begin{bmatrix} 0 & 0 \\ 1 & 0 \\ 0 & 1 \end{bmatrix} \mathbf{u}$$

has closed loop eigenvalues (−1, −2, −3).

$$\underline{\mathbf{y}} = \begin{bmatrix} 1 & 1 & 0 \\ 1 & 1 & 1 \end{bmatrix} \underline{\mathbf{x}}$$

Solution

We must first decide on the form of the matrix H.

Since (see (3.28))

$$H - A = BKC$$

and the first row of B is zero, it follows that the first row of

$$H - A$$

must be zero.

We must therefore choose H in the companion form.

Since the characteristic equation of H is

$$(\lambda+1)(\lambda+2)(\lambda+3) = \lambda^3 + 6\lambda^2 + 11\lambda + 6 = 0 .$$

$$H = \begin{bmatrix} 0 & 1 & 0 \\ 0 & 0 & 1 \\ -6 & -11 & -6 \end{bmatrix} \quad \text{(see (3.35))}$$

and hence (see (3.28))

$$Q = \begin{bmatrix} 0 & 0 & 0 \\ -3 & -3 & 0 \\ -8 & -8 & -8 \end{bmatrix} .$$

$$P = C' \otimes B = \begin{bmatrix} 1 & 1 \\ 1 & 1 \\ 0 & 1 \end{bmatrix} \otimes \begin{bmatrix} 0 & 0 \\ 1 & 0 \\ 0 & 1 \end{bmatrix} = \begin{bmatrix} 0 & 0 & 0 & 0 \\ 1 & 0 & 1 & 0 \\ 0 & 1 & 0 & 1 \\ 0 & 0 & 0 & 0 \\ 1 & 0 & 1 & 0 \\ 0 & 1 & 0 & 1 \\ 0 & 0 & 0 & 0 \\ 0 & 0 & 1 & 0 \\ 0 & 0 & 0 & 1 \end{bmatrix} .$$

An appropriate matrix T is the following

$$T = \begin{bmatrix} 0 & 0 & 0 & 0 & 0 & 0 & 0 & 1 & 0 \\ 0 & 0 & 0 & 0 & 0 & 0 & 0 & 0 & 1 \\ 0 & 1 & 0 & 0 & 0 & 0 & 0 & 0 & 0 \\ 0 & 0 & 1 & 0 & 0 & 0 & 0 & 0 & 0 \\ 1 & 0 & 0 & 0 & 0 & 0 & 0 & 0 & 0 \\ 0 & 0 & 0 & 1 & 0 & 0 & 0 & 0 & 0 \\ 0 & 0 & 0 & 0 & 0 & 0 & 1 & 0 & 0 \\ 0 & 0 & 1 & 0 & 0 & -1 & 0 & 0 & 0 \\ 0 & 1 & 0 & 0 & -1 & 0 & 0 & 0 & 0 \end{bmatrix} .$$

It follows that

$$TP = \left[\begin{array}{cccc} 0 & 0 & 1 & 0 \\ 0 & 0 & 0 & 1 \\ 1 & 0 & 1 & 0 \\ 0 & 1 & 0 & 1 \\ \hdashline 0 & 0 & 0 & 0 \\ 0 & 0 & 0 & 0 \\ 0 & 0 & 0 & 0 \\ 0 & 0 & 0 & 0 \\ 0 & 0 & 0 & 0 \end{array}\right] = \left[\begin{array}{c} P_1 \\ \hdashline P_2 \end{array}\right]$$

and

$$T\mathbf{q} = \left[\begin{array}{c} 0 \\ -8 \\ -3 \\ -8 \\ \hdashline 0 \\ 0 \\ 0 \\ 0 \\ 0 \end{array}\right] = \left[\begin{array}{c} \mathbf{u} \\ \hdashline \mathbf{v} \end{array}\right] .$$

Since

$$P_1 = \begin{bmatrix} 0 & 0 & 1 & 0 \\ 0 & 0 & 0 & 1 \\ 1 & 0 & 1 & 0 \\ 0 & 1 & 0 & 1 \end{bmatrix}, \quad P_1^{-1} = \begin{bmatrix} -1 & 0 & 1 & 0 \\ 0 & -1 & 0 & 1 \\ 1 & 0 & 0 & 0 \\ 0 & 1 & 0 & 0 \end{bmatrix}$$

so that (see (3.34))

$$\mathbf{k} = P_1^{-1}\mathbf{u} = \begin{bmatrix} -3 \\ 0 \\ 0 \\ -8 \end{bmatrix} .$$

Hence

$$K = \begin{bmatrix} -3 & 0 \\ 0 & -8 \end{bmatrix}$$

CHAPTER 4

Introduction to Matrix Calculus

4.1 INTRODUCTION

It is becoming ever increasingly clear that there is a real need for matrix calculus in fields such as multivariate analysis. There is a strong analogy here with matrix algebra which is such a powerful and elegant tool in the study of linear systems and elsewhere.

Expressions in multivariate analysis can be written in terms of scalar calculus, but the compactness of the equivalent relations in terms of matrices not only leads to a better understanding of the problems involved, but also encourages the consideration of problems which may be too complex to tackle by scalar calculus.

We have already defined the derivative of a matrix with respect to a scalar (see (3.1)), we now generalise this concept. The process is frequently referred to as **formal** or **symbolic** matrix differentiation. The basic definitions involve the partial differentiation of scalar matrix functions with respect to all the elements of a matrix. These derivatives are the elements of a matrix, of the same order as the original matrix, which is defined as the derived matrix. The words 'formal' and 'symbolic' refer to the fact that the matrix derivatives are defined without the rigorous mathematical justification which we expect for the corresponding scalar derivatives. This is not to say that such justification cannot be made, rather the fact is that this topic is still in its infancy and that appropriate mathematical basis is being laid as the subject develops. With this in mind we make the following observations about the notation used. In general the elements of the matrices A, B, C, ... will be constant scalars. On the other hand the elements of the matrices X, Y, Z, ... are scalar variables and we exclude the possibility that any element can be a constant or zero. In general we will also demand that these elements are independent. When this is not the case, for example when the matrix X is symmetric, is considered as a special case. The reader will appreciate the necessity for these restrictions when he considers the partial derivatives of (say) a matrix X with respect to one of its elements x_{rs}. Obviously the derivative is undefined if x_{rs} is a constant. The derivative is E_{rs} if x_{rs} is independent of all the other elements of X, but is $E_{rs} + E_{sr}$ if X is symmetric.

There have been attempts to define the derivative when x_{rs} is a constant (or zero) but, as far as this author knows, no rigorous mathematical theory for the general case has been proposed and successfully applied.

4.2 THE DERIVATIVES OF VECTORS

Let **x** and **y** be vectors of orders n and m respectively. We can define various derivatives in the following way [15]:

(1) The derivative of the vector **y** with respect to vector **x** is the matrix

$$\frac{\partial \mathbf{y}}{\partial \mathbf{x}} = \begin{bmatrix} \frac{\partial y_1}{\partial x_1} & \frac{\partial y_2}{\partial x_1} & \cdots & \frac{\partial y_m}{\partial x_1} \\ \frac{\partial y_1}{\partial x_2} & \frac{\partial y_2}{\partial x_2} & \cdots & \frac{\partial y_m}{\partial x_2} \\ \vdots & \vdots & & \vdots \\ \frac{\partial y_1}{\partial x_n} & \frac{\partial y_2}{\partial x_n} & \cdots & \frac{\partial y_m}{\partial x_n} \end{bmatrix} \tag{4.1}$$

of order $(n \times m)$ where $y_1, y_2, \ldots, y_m$ and $x_1, x_2, \ldots, x_n$ are the components of **y** and **x** respectively.

(2) The derivatives of a scalar with respect to a vector. If y is a scalar

$$\frac{\partial y}{\partial \mathbf{x}} = \begin{bmatrix} \frac{\partial y}{\partial x_1} \\ \frac{\partial y}{\partial x_2} \\ \vdots \\ \frac{\partial y}{\partial x_n} \end{bmatrix} . \tag{4.2}$$

(3) The derivative of a vector **y** with respect to a scalar x

$$\frac{\partial \mathbf{y}}{\partial x} = \begin{bmatrix} \frac{\partial y_1}{\partial x} & \frac{\partial y_2}{\partial x} & \cdots & \frac{\partial y_m}{\partial x} \end{bmatrix} . \tag{4.3}$$

Example 4.1

Given

$$\mathbf{y} = \begin{bmatrix} y_1 \\ y_2 \end{bmatrix}, \quad \mathbf{x} = \begin{bmatrix} x_1 \\ x_2 \\ x_3 \end{bmatrix}$$

and

$$y_1 = x_1^2 - x_2$$
$$y_2 = x_3^2 + 3x_2$$

Obtain $\partial \mathbf{y}/\partial \mathbf{x}$.

Solution

$$\frac{\partial \mathbf{y}}{\partial \mathbf{x}} = \begin{bmatrix} \frac{\partial y_1}{\partial x_1} & \frac{\partial y_2}{\partial x_1} \\ \frac{\partial y_1}{\partial x_2} & \frac{\partial y_2}{\partial x_2} \\ \frac{\partial y_1}{\partial x_3} & \frac{\partial y_2}{\partial x_3} \end{bmatrix} = \begin{bmatrix} 2x_1 & 0 \\ -1 & 3 \\ 0 & 2x_3 \end{bmatrix} .$$

In multivariate analysis, if $\mathbf{x}$ and $\mathbf{y}$ are of the same order, the absolute value of the determinant of $\partial \mathbf{x}/\partial \mathbf{y}$, that is of

$$\left| \frac{\partial \mathbf{x}}{\partial \mathbf{y}} \right|$$

is called the Jacobian of the transformation determined by

$$\mathbf{y} = \mathbf{y}(\mathbf{x}) .$$

Example 4.2

The transformation from spherical to cartesian co-ordinates is defined by $x = r \sin \theta \cos \psi$, $y = r \sin \theta \sin \psi$, and $z = r \cos \theta$ where $r > 0$, $0 < \theta < \pi$ and $0 \leqslant \psi < 2\pi$.

Obtain the Jacobian of the transformation.

Solution

Let

$$x = y_1 , \quad y = x_2 , \quad z = x_3$$

and

$$r = y_1 , \quad \theta = y_2 , \quad \psi = y_3 ,$$

$$J = \left| \frac{\partial \mathbf{x}}{\partial \mathbf{y}} \right| = \begin{vmatrix} \sin y_2 \cos y_3 & \sin y_2 \sin y_3 & \cos y_2 \\ y_1 \cos y_2 \cos y_3 & y_1 \cos y_2 \sin y_3 & -y_1 \sin y_2 \\ -y_1 \sin y_2 \sin y_3 & y_1 \sin y_2 \cos y_3 & 0 \end{vmatrix}$$
$$= y_1^2 \sin y_2 .$$

Definitions (4.1), (4.2) and (4.3) can be used to obtain derivatives to many frequently used expressions, including quatratic and bilinear forms.

For example consider

$$y = \mathbf{x}'A\mathbf{x}$$

Using (4.2) it is not difficult to show that

$$\frac{\partial y}{\partial \mathbf{x}} = A\mathbf{x} + A'\mathbf{x}$$
$$= 2A\mathbf{x} \text{ if } A \text{ is symmetric.}$$

We can of course differentiate the vector $2A\mathbf{x}$ with respect to $\mathbf{x}$, by definition (4.1).

$$\frac{\partial}{\partial \mathbf{x}}\left(\frac{\partial y}{\partial \mathbf{x}}\right) = \frac{\partial}{\partial \mathbf{x}}(2A\mathbf{x})$$
$$= 2A' = 2A \text{ (if } A \text{ is symmetric).}$$

The following table summarises a number of vector derivative formulae.

y scalar or a vector	$\frac{\partial y}{\partial \mathbf{x}}$
$A\mathbf{x}$	A'
$\mathbf{x}'A$	A
$\mathbf{x}'\mathbf{x}$	$2\mathbf{x}$
$\mathbf{x}'A\mathbf{x}$	$A\mathbf{x} + A'\mathbf{x}$

(4.4)

4.3 THE CHAIN RULE FOR VECTORS

Let

$$\mathbf{x} = \begin{bmatrix} x_1 \\ x_2 \\ \vdots \\ x_n \end{bmatrix}, \quad \mathbf{y} = \begin{bmatrix} y_1 \\ y_2 \\ \vdots \\ y_r \end{bmatrix} \quad \text{and} \quad \mathbf{z} = \begin{bmatrix} z_1 \\ z_2 \\ \vdots \\ z_m \end{bmatrix}.$$

Using the definition (4.1), we can write

$$\left(\frac{\partial \mathbf{z}}{\partial \mathbf{x}}\right)' = \begin{bmatrix} \frac{\partial z_1}{\partial x_1} & \frac{\partial z_1}{\partial x_2} & \cdots & \frac{\partial z_1}{\partial x_n} \\ \frac{\partial z_2}{\partial x_1} & \frac{\partial z_2}{\partial x_2} & \cdots & \frac{\partial z_2}{\partial x_n} \\ \vdots & \vdots & & \vdots \\ \frac{\partial z_m}{\partial x_1} & \frac{\partial z_m}{\partial x_2} & \cdots & \frac{\partial z_m}{\partial x_n} \end{bmatrix}. \tag{4.5}$$

Assume that

$$\mathbf{z} = \mathbf{y}(\mathbf{x})$$

so that

$$\frac{\partial z_i}{\partial x_j} = \sum_{q=1}^{r} \frac{\partial z_i}{\partial y_q} \frac{\partial y_q}{\partial x_j} \qquad \begin{array}{l} i = 1, 2, \ldots, m \\ j = 1, 2, \ldots, n. \end{array}$$

Then (4.5) becomes

$$\left(\frac{\partial \mathbf{z}}{\partial \mathbf{x}}\right)' = \begin{bmatrix} \sum \frac{\partial z_1}{y_q} \frac{\partial y_q}{\partial x_1} & \sum \frac{\partial z_1}{\partial y_q} \frac{\partial y_q}{\partial x_2} & \cdots & \sum \frac{\partial z_1}{\partial y_q} \frac{\partial y_q}{\partial x_n} \\ \sum \frac{\partial z_2}{\partial y_q} \frac{\partial y_q}{\partial x_1} & \sum \frac{\partial z_2}{\partial y_q} \frac{\partial y_q}{\partial x_2} & \cdots & \sum \frac{\partial z_2}{\partial y_q} \frac{\partial y_q}{\partial x_n} \\ \vdots & & & \\ \sum \frac{\partial z_m}{\partial y_q} \frac{\partial y_q}{\partial x_1} & \sum \frac{\partial z_m}{\partial y_q} \frac{\partial y_q}{\partial x_2} & \cdots & \sum \frac{\partial z_m}{\partial y_q} \frac{\partial y_q}{\partial x_n} \end{bmatrix}$$

$$= \begin{bmatrix} \frac{\partial z_1}{\partial y_1} & \frac{\partial z_1}{\partial y_2} & \cdots & \frac{\partial z_1}{\partial y_r} \\ \frac{\partial z_2}{\partial y_1} & \frac{\partial z_2}{\partial y_2} & \cdots & \frac{\partial z_2}{\partial y_r} \\ \vdots & & & \\ \frac{\partial z_m}{\partial y_1} & \frac{\partial z_m}{\partial y_2} & \cdots & \frac{\partial z_m}{\partial y_r} \end{bmatrix} \begin{bmatrix} \frac{\partial y_1}{\partial x_1} & \frac{\partial y_1}{\partial x_2} & \cdots & \frac{\partial y_1}{\partial x_n} \\ \frac{\partial y_2}{\partial x_1} & \frac{\partial y_2}{\partial x_2} & \cdots & \frac{\partial y_2}{\partial x_n} \\ \vdots & & & \\ \frac{\partial y_r}{\partial x_1} & \frac{\partial y_r}{\partial x_2} & \cdots & \frac{\partial y_r}{\partial x_n} \end{bmatrix}$$

$$= \left(\frac{\partial \mathbf{z}}{\partial \mathbf{y}}\right)' \left(\frac{\partial \mathbf{y}}{\partial \mathbf{x}}\right)' \qquad \text{(by (4.1))}$$

$$= \left(\frac{\partial \mathbf{y}}{\partial \mathbf{x}} \frac{\partial \mathbf{z}}{\partial \mathbf{y}}\right)'$$

on transposing both sides, we finally obtain

$$\frac{\partial \mathbf{z}}{\partial \mathbf{y}} = \frac{\partial \mathbf{y}}{\partial \mathbf{x}} \frac{\partial \mathbf{z}}{\partial \mathbf{y}} \quad . \tag{4.6}$$

4.4 THE DERIVATIVE OF SCALAR FUNCTIONS OF A MATRIX WITH RESPECT TO THE MATRIX

Let $X = [x_{ij}]$ be a matrix of order $(m \times n)$ and let

$$y = f(X)$$

be a scalar function of X.

The derivative of y with respect to X, denoted by

$$\frac{\partial y}{\partial X}$$

is defined as the following matrix of order $(m \times n)$

$$\frac{\partial y}{\partial X} = \begin{bmatrix} \frac{\partial y}{\partial x_{11}} & \frac{\partial y}{\partial x_{12}} & \cdots & \frac{\partial y}{\partial x_{1n}} \\ \frac{\partial y}{\partial x_{21}} & \frac{\partial y}{\partial x_{22}} & \cdots & \frac{\partial y}{\partial x_{2n}} \\ \vdots & \vdots & & \vdots \\ \frac{\partial y}{\partial x_{m1}} & \frac{\partial y}{\partial x_{m2}} & \cdots & \frac{\partial y}{\partial x_{mn}} \end{bmatrix} = \left[\frac{\partial y}{\partial x_{ij}}\right] = \sum_{i,j} E_{ij} \frac{\partial y}{\partial x_{ij}} \tag{4.7}$$

where E_{ij} is an elementary matrix of order $(m \times n)$.

Definition

When $X = [x_{ij}]$ is a matrix of order $(m \times n)$ and $y = f(X)$ is a scalar function of X, then $\partial f(X)/\partial X$ is known as a **gradient matrix**.

Example 4.3

Given the matrix $X = [x_{ij}]$ of order $(n \times n)$ obtain $\partial y/\partial X$ when $y = \text{tr } X$.

Solution

$y = \text{tr } X = x_{11} + x_{22} + \ldots + x_{nn} = \text{tr } X'$ (see 1.33) hence by (4.7)

$$\frac{\partial y}{\partial X} = I_n .$$

An important family of derivatives with respect to a matrix involves functions of the determinant of a matrix, for example

$$y = |X| \quad \text{or} \quad y = |AX| .$$

We will consider a general case, say we have a matrix $Y = [y_{ij}]$ whose components are functions of a matrix $X = [x_{ij}]$, that is

$$y_{ij} = f_{ij}(\mathbf{x})$$

where $\mathbf{x} = [x_{11}\, x_{12} \ldots x_{mn}]'$.

We will determine

$$\frac{\partial |Y|}{\partial x_{rs}}$$

which will allow us to build up the matrix

$$\frac{\partial |Y|}{\partial X} \, .$$

Using the chain rule we can write

$$\frac{\partial |Y|}{\partial x_{rs}} = \sum_i \sum_j \frac{\partial |Y|}{\partial y_{ij}} \cdot \frac{\partial y_{ij}}{\partial x_{rs}} \, .$$

But $|Y| = \sum_j y_{ij} Y_{ij}$

where Y_{ij} is the cofactor of the element y_{ij} in $|Y|$. Since the cofactors $Y_{i1}, Y_{i2}, \ldots$ are independent of the element y_{ij}, we have

$$\frac{\partial |Y|}{\partial y_{ij}} = Y_{ij} \, .$$

It follows that

$$\frac{\partial |Y|}{\partial x_{rs}} = \sum_i \sum_j Y_{ij} \frac{\partial y_{ij}}{\partial x_{rs}} \, . \tag{4.8}$$

Although we have achieved our objective in determining the above formula, it can be written in an alternate and useful form.

With

$$a_{ij} = Y_{ij} \quad \text{and} \quad b_{ij} = \frac{\partial y_{ij}}{\partial x_{rs}}$$

we can write (4.8) as

$$\begin{aligned}
\frac{\partial |Y|}{\partial x_{rs}} &= \sum_i \sum_j a_{ij} b_{ij} = \sum_i \sum_j a_{ij} b_{ij} \mathbf{e}_j' \mathbf{e}_j \\
&= \sum_i \sum_j a_{ij} \mathbf{e}_j' b_{ij} \mathbf{e}_j \\
&= \sum_i A_{i\cdot}' B_{i\cdot} \qquad \text{(by (1.23) and (1.24))} \\
&= \operatorname{tr}(AB') = \operatorname{tr}(B'A) \quad \text{(by (1.43))}
\end{aligned}$$

where $A = [a_{ij}]$ and $B = [b_{ij}]$.

Assuming that Y is of order $(k \times k)$ let

$$\begin{bmatrix} Y_{11} & Y_{12} & \dots & Y_{1k} \\ Y_{21} & Y_{22} & \dots & Y_{2k} \\ \vdots & & & \\ Y_{k1} & Y_{k2} & \dots & Y_{kk} \end{bmatrix} = Z \tag{4.9}$$

and since

$$\left[\frac{\partial y_{ij}}{\partial x_{rs}}\right] = \frac{\partial Y}{\partial x_{rs}}$$

we can write

$$\frac{\partial |Y|}{\partial x_{rs}} = \operatorname{tr}\left(\frac{\partial Y'}{\partial x_{rs}} Z\right) . \tag{4.10}$$

We use (4.10) to evaluate $\partial|Y|/\partial x_{11}$, $\partial|Y|/\partial x_{12}$, ..., $\partial|Y|/\partial x_{mn}$ and then use (4.7) to construct

$$\frac{\partial |Y|}{\partial X} ,$$

Example 4.4

Given the matrix $X = [x_{ij}]$ of order (2×2) evaluate $\partial|X|/\partial X$,

(i) when all the components x_{ij} of X are independent
(ii) when X is a symmetric matrix.

Solution

(i) In the notation of (4.10), we have

$$Y = \begin{bmatrix} x_{11} & x_{12} \\ x_{21} & x_{22} \end{bmatrix}$$

so that $\partial Y/\partial x_{rs} = E_{rs}$ (for notation see (1.4)).

As

$$Z = \begin{bmatrix} X_{11} & X_{12} \\ X_{21} & X_{22} \end{bmatrix} ,$$

we use the result of Example (1.4) to write (4.10) as

$$\frac{\partial |Y|}{\partial x_{rs}} = (\operatorname{vec} E_{rs})' \operatorname{vec} Z .$$

So that, for example

$$\frac{\partial|Y|}{\partial x_{11}} = [1\ 0\ 0\ 0]\begin{bmatrix} X_{11} \\ X_{21} \\ X_{12} \\ X_{22} \end{bmatrix} = X_{11}$$

and

$$\frac{\partial|Y|}{\partial x_{12}} = [0\ 0\ 1\ 0]\begin{bmatrix} X_{11} \\ X_{21} \\ X_{12} \\ X_{22} \end{bmatrix} = X_{12} \quad \text{and so on.}$$

Hence

$$\frac{\partial|Y|}{\partial X} = \frac{\partial|X|}{\partial X} = \begin{bmatrix} X_{11} & X_{12} \\ X_{21} & X_{22} \end{bmatrix}$$

$$= |X|(X^{-1})' \qquad \text{(See [18] p. 124).}$$

(ii) This time

$$Y = \begin{bmatrix} x_{11} & x_{12} \\ x_{12} & x_{22} \end{bmatrix}$$

hence

$$\frac{\partial|Y|}{x_{11}} = E_{11}\,, \quad \frac{\partial|Y|}{x_{12}} = E_{12} + E_{21} \quad \text{and so on.}$$

(See the introduction to Chapter 4 for explanantion of the notation.)

It follows that

$$\frac{\partial|Y|}{\partial x_{12}} = \frac{\partial|Y|}{\partial x_{21}} = [0\ 1\ 1\ 0]\begin{bmatrix} X_{11} \\ X_{21} \\ X_{12} \\ X_{22} \end{bmatrix} = X_{21} + X_{12} = 2X_{12}$$

$$\text{(Since } X_{12} = X_{21}\text{)}$$

hence

$$\frac{\partial|Y|}{\partial X} = \begin{bmatrix} X_{11} & 2X_{12} \\ 2X_{21} & X_{22} \end{bmatrix} = 2\begin{bmatrix} X_{11} & X_{12} \\ X_{21} & X_{22} \end{bmatrix} - \begin{bmatrix} X_{11} & 0 \\ 0 & X_{22} \end{bmatrix}.$$

The above results can be generalised to a matrix X of order $(n \times n)$.

We obtain, in the symmetric matrix case

$$\frac{\partial|X|}{\partial X} = 2[X_{ij}] - \operatorname{diag}\{X_{ii}\}\,.$$

We defer the discussion of differentiating other scalar matrix functions to Chapter 5.

4.5 THE DERIVATIVE OF A MATRIX WITH RESPECT TO ONE OF ITS ELEMENTS AND CONVERSELY

In this section we will generalise the concepts discussed in the previous section. We again consider a matrix

$$X = [x_{ij}] \text{ or order } (m \times n) .$$

The derivative of the matrix X relative to one of its elements x_{rs} (say), is obviously (see (3.1))

$$\frac{\partial X}{\partial x_{rs}} = E_{rs} \tag{4.11}$$

where E_{rs} is the elementary matrix *of order $(m \times n)$ (the order of X)* defined in section 1.2.

It follows immediately that

$$\frac{\partial X'}{\partial x_{rs}} = E'_{rs} . \tag{4.12}$$

A more complicated situation arises when we consider a product of the form

$$Y = AXB \tag{4.13}$$

where

$$X = [x_{ij}] \text{ is of order } (m \times n)$$

$$A = [a_{ij}] \text{ is or order } (l \times m)$$

$$B = [b_{ij}] \text{ is of order } (n \times q)$$

and

$$Y = [y_{ij}] \text{ is of order } (l \times q) .$$

A and B are assumed independent of X.

Our aim is to find the rule for obtaining the derivatives

$$\frac{\partial Y}{\partial x_{rs}} \quad \text{and} \quad \frac{\partial y_{ij}}{\partial X}$$

where x_{rs} is a typical element of X and y_{ij} is a typical element of Y.

We will first obtain the (i,j)th element y_{ij} in (4.13) as a function of the elements of X.

We can achieve this objective in a number of different ways. For example, we can use (2.13) to write

$$\text{vec}\, Y = (B' \otimes A)\, \text{vec}\, X .$$

From this expression we see that y_{ij} is the (scalar) product of the ith row of

$$[b_{1j}A \vdots b_{2j}A \vdots \ldots \vdots b_{nj}A] \quad \text{and vec } X,$$

so that

$$y_{ij} = \sum_{p=1}^{n} \sum_{l=1}^{m} a_{il} b_{pj} x_{lp} \quad . \tag{4.14}$$

From (4.14) we immediately obtain

$$\frac{\partial y_{ij}}{\partial x_{rs}} = a_{ir} b_{sj} \quad . \tag{4.15}$$

We can now write the expression for $\partial y_{ij}/\partial X$,

$$\frac{\partial y_{ij}}{\partial X} = \begin{bmatrix} \frac{\partial y_{ij}}{\partial x_{11}} & \frac{\partial y_{ij}}{\partial x_{12}} & \cdots & \frac{\partial y_{ij}}{\partial x_{1n}} \\ \frac{\partial y_{ij}}{\partial x_{21}} & \frac{\partial y_{ij}}{\partial x_{22}} & \cdots & \frac{\partial y_{ij}}{\partial x_{2n}} \\ \vdots & & & \\ \frac{\partial y_{ij}}{\partial x_{m1}} & \frac{\partial y_{ij}}{\partial x_{m2}} & \cdots & \frac{\partial y_{ij}}{\partial x_{mn}} \end{bmatrix} . \tag{4.16}$$

Using (4.15), we obtain

$$\frac{\partial y_{ij}}{\partial X} = \begin{bmatrix} a_{i1}b_{1j} & a_{i1}b_{2j} & \cdots & a_{i1}b_{nj} \\ a_{i2}b_{1j} & a_{i2}b_{2j} & \cdots & a_{i2}b_{nj} \\ \vdots & & & \\ a_{im}b_{1j} & a_{im}b_{2j} & \cdots & a_{im}b_{nj} \end{bmatrix} . \tag{4.17}$$

We note that the matrix on the right hand side of (4.17) can be expressed as (for notation see (1.5)(1.13)(1.16) and (1.17))

$$\begin{aligned} \frac{\partial y_{ij}}{\partial X} &= \begin{bmatrix} a_{i1} \\ a_{i2} \\ \vdots \\ a_{im} \end{bmatrix} [b_{1j} b_{2j} \ldots b_{nj}] \\ &= A_{i\cdot} B_{\cdot j}{}' \\ &= A' \mathbf{e}_i \mathbf{e}_j' B' . \end{aligned}$$

So that

$$\frac{\partial y_{ij}}{\partial X} = A'E_{ij}B' \tag{4.18}$$

where E_{ij} is an elementary matrix of order $(l \times q)$ *the order of the matrix* Y.

We also use (4.14) to obtain an expression for $\partial Y/\partial x_{rs}$.

$$\frac{\partial Y}{\partial x_{rs}} = \left[\frac{\partial y_{ij}}{\partial x_{rs}}\right] \qquad (r, s \text{ fixed}, i, j \text{ variable } 1 \leqslant i \leqslant l, 1 \leqslant j \leqslant q)$$

that is

$$\frac{\partial Y}{\partial x_{rs}} = \begin{bmatrix} \frac{\partial y_{11}}{\partial x_{rs}} & \frac{\partial y_{12}}{\partial x_{rs}} & \cdots & \frac{\partial y_{1q}}{\partial x_{rs}} \\ \frac{\partial y_{21}}{\partial x_{rs}} & \frac{\partial y_{22}}{\partial x_{rs}} & \cdots & \frac{\partial y_{2q}}{\partial x_{rs}} \\ \vdots & & & \\ \frac{\partial y_{l1}}{\partial x_{rs}} & \frac{\partial y_{l2}}{\partial x_{rs}} & \cdots & \frac{\partial y_{lq}}{\partial x_{rs}} \end{bmatrix} = \sum_{i,j} E_{ij} \frac{\partial y_{ij}}{\partial x_{rs}} \tag{4.19}$$

where E_{ij} is an elementary matrix of order $(l \times q)$.

We again use (4.15) to write

$$\left[\frac{\partial y_{ij}}{\partial x_{rs}}\right] = \begin{bmatrix} a_{1r}b_{s1} & a_{1r}b_{s2} & \cdots & a_{1r}b_{sq} \\ a_{2r}b_{s1} & a_{2r}b_{s2} & \cdots & a_{2r}b_{sq} \\ \vdots & & & \\ a_{mr}b_{s1} & a_{mr}b_{s2} & \cdots & a_{mr}b_{sq} \end{bmatrix}$$

$$= \begin{bmatrix} a_{1r} \\ a_{2r} \\ \vdots \\ a_{mr} \end{bmatrix} [b_{s1}\ b_{s2} \ldots b_{sq}]$$

$$= A_{\cdot r}B_{s\cdot}' = Ae_r e_s' B\ .$$

So that

$$\frac{\partial(AXB)}{\partial x_{rs}} = AE_{rs}B \tag{4.20}$$

where E_{rs} is an elementary matrix of order $(m \times n)$, *the order of the matrix* X.

Example 4.5

Find the derivative $\partial Y/\partial x_{rs}$, given

$$Y = AX'B$$

where the order of the matrices A, X and B is such that the product on the right hand side is defined.

Solution

By the method used above to obtain the derivative $\partial/\partial x_{rs}\,(AXB)$, we find

$$\frac{\partial}{\partial x_{rs}}\,(AX'B) = AE'_{rs}B\ .$$

Before continuing with further examples we need a rule for determining the derivative of a product of matrices.

Consider

$$Y = UV \tag{4.21}$$

where $U = [u_{ij}]$ is of order $(m \times n)$ and $V = [v_{ij}]$ is of order $(n \times l)$ and both U and V are functions of a matrix X.

We wish to determine

$$\frac{\partial Y}{\partial x_{rs}} \quad \text{and} \quad \frac{\partial y_{ij}}{\partial X}\ .$$

The (i,j)th element of (4.21) is

$$y_{ij} = \sum_{p=1}^{n} u_{ip}v_{pj} \tag{4.22}$$

hence

$$\frac{\partial y_{ij}}{\partial x_{rs}} = \sum_{p=1}^{n} \frac{\partial u_{ip}}{\partial x_{rs}}v_{pj} + \sum_{p=1}^{n} u_{ip}\frac{\partial v_{pj}}{\partial x_{rs}}\ . \tag{4.23}$$

For fixed r and s, (4.23) is the (i,j)th element of the matrix $\partial Y/\partial x_{rs}$ of order $(m \times l)$ the same as the order of the matrix Y.

On comparing both the terms on the right hand side of (4.23) with (4.22), we can write

$$\frac{\partial(UV)}{\partial x_{rs}} = \frac{\partial U}{\partial x_{rs}}V + U\frac{\partial V}{\partial x_{rs}} \tag{4.24}$$

as one would expect.

On the other hand, when fixing (i,j), (4.23) is the (r,s)th element of the matrix $\partial y_{ij}/\partial X$, which is of the same order as the matrix X, that is

$$\frac{\partial y_{ij}}{\partial X} = \sum_{p=1}^{n} \frac{\partial u_{ip}}{\partial X} v_{pj} + \sum_{p=1}^{n} u_{ip} \frac{\partial v_{pj}}{\partial X} \tag{4.25}$$

We will make use of the result (4.24) in some of the subsequent examples.

Example 4.6

Let $X = [x_{rs}]$ be a non-singular matrix. Find the derivative $\partial Y/\partial x_{rs}$, given

(i) $Y = AX^{-1}B$, and
(ii) $Y = X'AX$

Solution

(i) Using (4.24) to differentiate

$$YY^{-1} = I\,,$$

we obtain

$$\frac{\partial Y}{\partial x_{rs}} Y^{-1} + Y \frac{\partial Y^{-1}}{\partial x_{rs}} = 0\,,$$

hence

$$\frac{\partial Y}{\partial x_{rs}} = -Y \frac{\partial Y^{-1}}{\partial x_{rs}} Y^{-}\,.$$

But by (4.20)

$$\frac{\partial Y^{-1}}{\partial x_{rs}} = \frac{\partial}{\partial x_{rs}}(B^{-1}XA^{-1}) = B^{-1}E_{rs}A^{-1}$$

so that

$$\frac{\partial Y}{\partial x_{rs}} = \frac{\partial}{\partial x_{rs}}(AX^{-1}B) = -AX^{-1}BB^{-1}E_{rs}A^{-1}AX^{-1}B$$
$$= -AX^{-1}E_{rs}X^{-1}B\,.$$

(ii) Using (4.24), we obtain

$$\frac{\partial Y}{\partial x_{rs}} = \frac{\partial X'}{\partial x_{rs}} AX + X' \frac{\partial (AX)}{\partial x_{rs}}$$
$$= E'_{rs}AX + X'AE_{rs} \qquad \text{(by (4.12) and (4.20))}\,.$$

Both (4.18) and (4.20) were derived from (4.15) which is valid for all i, j and r, s, defined by the orders of the matrices involved.

The First Transformation Principle

It follows that (4.18) is a transformation of (4.20) and conversely. To obtain (4.18) from (4.20) we replace A by A', B by B' and E_{rs} by E_{ij} (careful, E_{rs} and E_{ij} may be of different orders).

The interesting point is that although (4.18) and (4.20) were derived for constant matrices A and B, the above transformation is independent of the status of the matrices and is valid even when A and B are functions of X.

Example 4.7

Find the derivative of $\partial y_{ij}/\partial X$, given

(i) $Y = AX'B$,
(ii) $Y = AX^{-1}B$, and
(iii) $Y = X'AX$

where $X = [x_{ij}]$ is a nonsingular matrix.

Solution

(i) Let $W = X'$, then

$$Y = AWB \text{ so that by (4.20) } \frac{\partial Y}{\partial w_{rs}} = AE_{rs}B$$

hence

$$\frac{\partial y_{ij}}{\partial W} = A'E_{ij}B'.$$

But

$$\frac{\partial y_{ij}}{\partial X} = \frac{\partial y_{ij}}{\partial W'} = \left(\frac{\partial y_{ij}}{\partial W}\right)'$$

hence

$$\frac{\partial y_{ij}}{\partial X} = BE'_{ij}A$$

(ii) From Example 4.6(i)

$$\frac{\partial Y}{\partial x_{rs}} = -AX^{-1}E_{rs}X^{-1}B.$$

Let $A_1 = AX^{-1}$ and $B_1 = X^{-1}B$, then

$$\frac{\partial Y}{\partial x_{rs}} = -A_1E_{rs}B_1$$

so that

$$\frac{\partial y_{ij}}{\partial X} = -A_1'E_{ij}B_1' = -(X^{-1})'A'E_{ij}B'(X^{-1})'.$$

(iii) From Example 4.6(ii)

$$\frac{\partial Y}{\partial x_{rs}} = E'_{rs}AX + X'AE_{rs} .$$

Let $A_1 = I, B_1 = AX, A_2 = X'A$ and $B_2 = I$, then

$$\frac{\partial Y}{\partial x_{rs}} = A_1E'_{rs}B_1 + A_2E_{rs}B_2 .$$

The second term on the right hand side is in standard form. The first term is in the form of the solution to Example 4.5 for which the derivative $\partial y_{ij}/\partial X$ was found in (i) above, hence

$$\frac{\partial y_{ij}}{\partial X} = B_1E'_{ij}A_1 + A'_2E_{ij}B'_2$$

$$= AXE'_{ij} + A'XE_{ij} .$$

It is interesting to compare this last result with the example in section 4.2 when we considered the scalar $y = \mathbf{x}'A\mathbf{x}$.

In this special case when the matrix X has only one column, the elementary matrix which is of the same order as Y, becomes

$$E_{ij} = E'_{ij} = 1 ,$$

Hence

$$\frac{\partial y_{ij}}{\partial X} = \frac{\partial y}{\partial \mathbf{x}} = A\mathbf{x} + A'\mathbf{x}$$

which is the result obtained in section 4.2 (see (4.4)).

Conversely using the above techniques we can also obtain the derivatives of the matrix equivalents of the other equations in the table (4.4).

Example 4.8

Find

$$\frac{\partial Y}{\partial x_{rs}} \quad \text{and} \quad \frac{\partial y_{ij}}{\partial X}$$

when

(i) $Y = AX$, and
(ii) $Y = X'X$.

Solution

(i) With $B = I$, apply (4.20)

$$\frac{\partial Y}{\partial x_{rs}} = AE_{rs} .$$

The transformation principle results in

$$\frac{\partial y_{ij}}{\partial X} = A'E_{ij} .$$

(ii) This is a special case of Example 4.6(ii) in which $A = I$.
We have found the solution

$$\frac{\partial Y}{\partial x_{rs}} = E'_{rs}X + X'E_{rs}$$

and (Solution to Example 4.7(iii))

$$\frac{\partial y_{ij}}{\partial X} = XE'_{ij} + XE_{ij} .$$

4.6 THE DERIVATIVES OF THE POWERS OF A MATRIX

Our aim in this section is to obtain the rules for determining

$$\frac{\partial Y}{\partial x_{rs}} \quad \text{and} \quad \frac{\partial y_{ij}}{\partial X}$$

when

$$Y = X^n.$$

Using (4.24) when $U = V = X$ so that

$$Y = X^2$$

we immediately obtain

$$\frac{\partial Y}{\partial x_{rs}} = E_{rs}X + XE_{rs}$$

and, applying the first transformation principle,

$$\frac{\partial y_{ij}}{\partial X} = E_{ij}X' + X'E_{ij} .$$

It is instructive to repeat this exercise with

$$U = X^2 \quad \text{and} \quad V = X$$

so that

$$Y = X^3.$$

We obtain

$$\frac{\partial Y}{\partial x_{rs}} = E_{rs}X^2 + XE_{rs}X + X^2E_{rs}$$

and

$$\frac{\partial y_{ij}}{\partial X} = E_{ij}(X')^2 + X'E_{ij}X' + (X')^2E_{ij} .$$

More generally, it can be proved by induction, that for

$$Y = X^n$$

$$\frac{\partial Y}{\partial x_{rs}} = \sum_{k=0}^{n-1} X^k E_{rs} X^{n-k-1} \tag{4.26}$$

where by definition $X^0 = I$, and

$$\frac{\partial y_{ij}}{\partial X} = \sum_{k=1}^{n-1} (X')^k E_{ij} (X')^{n-k-1} \tag{4.27}$$

Example 4.9

Using the result (4.26), obtain $\partial Y/\partial x_{rs}$ when

$$Y = X^{-n}$$

Solution

Using (4.24) on both sides of

$$X^{-n}X^n = I$$

we find

$$\frac{\partial (X^{-n})}{\partial x_{rs}} X^n + X^{-n}\frac{\partial (X^n)}{\partial x_{rs}} = 0$$

so that

$$\frac{\partial (X^{-n})}{\partial x_{rs}} = -X^{-n}\frac{\partial (X^n)}{\partial x_{rs}} X^{-n}.$$

Now making use of (4.26), we conclude that

$$\frac{\partial (X^{-n})}{\partial x_{rs}} = -X^{-n}\left[\sum_{k=0}^{n-1} X^k E_{rs} X^{n-k-1}\right] X^{-n}.$$

Problems for Chapter 4

(1) Given

$$X = \begin{bmatrix} x_{11} & x_{12} & x_{13} \\ x_{21} & x_{22} & x_{23} \end{bmatrix}, \qquad Y = \begin{bmatrix} x & e^{2x} \\ 1 & x^{-1} \\ 2x^2 & \sin x \end{bmatrix}$$

and $y = 2x_{11}x_{22} - x_{21}x_{13}$, calculate

$$\frac{\partial y}{\partial X} \quad \text{and} \quad \frac{\partial Y}{\partial x}.$$

(2) Given

$$X = \begin{bmatrix} \sin x & x \\ \cos x & e^x \end{bmatrix} \quad \text{and} \quad X = \begin{bmatrix} \sin x & e^x \\ e^x & x \end{bmatrix}$$

evaluate

$$\frac{\partial |X|}{\partial X}$$

by

(a) a direct method
(b) use of a derivative formula.

(3) Given

$$X = \begin{bmatrix} x_{11} & x_{12} & x_{13} \\ x_{21} & x_{22} & x_{23} \end{bmatrix} \quad \text{and} \quad Y = X'X\,,$$

use a direct method to evaluate

$$\text{(a)}\ \frac{\partial Y}{\partial x_{21}} \quad \text{and} \quad \text{(b)}\ \frac{\partial y_{13}}{\partial X}$$

(4) Obtain expressions for

$$\frac{\partial Y}{\partial x_{rs}} \quad \text{and} \quad \frac{\partial y_{ij}}{\partial X}$$

when

(a) $Y = XAX$ and (b) $Y = X'AX'$.

(5) Obtain an expression for $\partial |AXB|/\partial x_{rs}$. It is assumed AXB is non-singular.

(6) Evaluate $\partial Y/\partial x_{rs}$ when

(a) $Y = X(X')^2$ and (b) $Y = (X')^2X$.

CHAPTER 5

Further Development of Matrix Calculus including an Application of Kronecker Products

5.1 INTRODUCTION

In Chapter 4 we discussed rules for determining the derivatives of a vector and then the derivatives of a matrix.

But it will be remembered that when Y is a matrix, then vec Y is a vector. This fact, together with the closely related Kronecker product techniques discussed in Chapter 2 will now be exploited to derive some interesting results.

Also we explore further the derivatives of some scalar functions with respect to a matrix first considered in the previous chapter.

5.2 DERIVATIVES OF MATRICES AND KRONECKER PRODUCTS

In the previous chapter we have found $\partial y_{ij}/\partial X$ when

$$Y = AXB \tag{5.1}$$

where $Y = [y_{ij}]$, $A = [a_{ij}]$, $X = [x_{ij}]$ and $B = [b_{ij}]$.

We now obtain $(\partial \operatorname{vec} Y)/(\partial \operatorname{vec} X)$ for (5.1). We can write (5.1) as

$$\mathbf{y} = P\mathbf{x} \tag{5.2}$$

where $\mathbf{y} = \operatorname{vec} Y$, $\mathbf{x} = \operatorname{vec} X$ and $P = B' \otimes A$.

By (4.1), (4.4) and (2.10)

$$\frac{\partial \mathbf{y}}{\partial \mathbf{x}} = P' = (B' \otimes A)' = B \otimes A'. \tag{5.3}$$

The corresponding result for the equation

$$Y = AX'B \tag{5.4}$$

is not so simple.

The problem is that when we write (5.4) in the form of (5.2), we have this time

$$\mathbf{y} = P\mathbf{z} \tag{5.5}$$

where $\mathbf{z} = \text{vec}\, X'$

We can find (see (2.25)) a permutation matrix U such that

$$\text{vec}\, X' = U \,\text{vec}\, X \tag{5.6}$$

in which case (5.5) becomes

$$\mathbf{y} = PU\mathbf{x}$$

so that

$$\frac{\partial \mathbf{y}}{\partial \mathbf{x}} = (PU)' = U'(B \otimes A') . \tag{5.7}$$

It is convenient to write

$$U'(B \otimes A') = (B \otimes A')_{(n)} \tag{5.8}$$

U' is seen to premultiply the matrix $(B \otimes A')$. Its effect is therefore to rearrange the rows of $(B \otimes A')$.

In fact the first and every subsequent nth row of $(B \otimes A')$ form the first consecutive m rows of $(B \otimes A')_{(n)}$. The second and every subsequent nth row form the next m consecutive rows of $(B \otimes A')_{(n)}$ and so on.

A special case of this notation is for $n = 1$, then

$$(B \otimes A')_{(1)} = B \otimes A' . \tag{5.9}$$

Now, returning to (5.5), we obtain, by comparison with (5.3)

$$\frac{\partial \mathbf{y}}{\partial \mathbf{x}} = (B \otimes A')_{(n)} \tag{5.10}$$

Example 5.1

Obtain $(\partial\, \text{vec}\, Y)/(\partial\, \text{vec}\, X)$, given $X = [x_{ij}]$ of order $(m \times n)$, when

(i) $Y = AX$, (ii) $Y = XA$, (iii) $Y = AX'$ and (iv) $Y = X'A$.

Solution

Let $\mathbf{y} = \text{vec}\, Y$ and $\mathbf{x} = \text{vec}\, X$.

(i) Use (5.3) with $B = I$

$$\frac{\partial \mathbf{y}}{\partial \mathbf{x}} = I \otimes A' .$$

(ii) Use (5.3)

$$\frac{\partial \mathbf{y}}{\partial \mathbf{x}} = A \otimes I \ .$$

(iii) Use (5.10)

$$\frac{\partial \mathbf{y}}{\partial \mathbf{x}} = (I \otimes A')_{(n)} \ .$$

(iv) Use (5.10)

$$\frac{\partial \mathbf{y}}{\partial \mathbf{x}} = (A \otimes I)_{(n)} \ .$$

5.3 THE DETERMINATION OF $(\partial \operatorname{vec} X)/(\partial \operatorname{vec} Y)$ FOR MORE COMPLICATED EQUATIONS

In this section we wish to determine the derivative $(\partial \operatorname{vec} Y)/(\partial \operatorname{vec} X)$ when, for example,

$$Y = X'AX \tag{5.11}$$

where X is of order $(m \times n)$.

Since Y is a matrix of order $(n \times n)$, it follows that vec Y and vec X are vectors of order nn and nm respectively.

With the usual notation

$$Y = [y_{ij}] \ , \quad X = [x_{ij}]$$

we have, by definition (4.1),

$$\frac{\partial \operatorname{vec} Y}{\partial \operatorname{vec} X} = \begin{bmatrix} \frac{\partial y_{11}}{\partial x_{11}} & \frac{\partial y_{21}}{\partial x_{11}} & \cdots & \frac{\partial y_{nn}}{\partial x_{11}} \\ \frac{\partial y_{11}}{\partial x_{21}} & \frac{\partial y_{21}}{\partial x_{21}} & \cdots & \frac{\partial y_{nn}}{\partial x_{21}} \\ \vdots & \vdots & & \vdots \\ \frac{\partial y_{11}}{\partial x_{mn}} & \frac{\partial y_{21}}{\partial x_{mn}} & \cdots & \frac{\partial y_{nn}}{\partial x_{mn}} \end{bmatrix} . \tag{5.12}$$

But by definition (4.19),

the first row of the matrix (5.12) is $\left(\operatorname{vec} \frac{\partial Y}{\partial x_{11}}\right)'$,

the second row of the matrix (5.12) is $\left(\operatorname{vec} \frac{\partial Y}{\partial x_{21}}\right)'$, etc.

We can therefore write (5.12) as

$$\frac{\partial \text{ vec } Y}{\partial \text{ vec } X} = \left[\text{vec } \frac{\partial Y}{\partial x_{11}} \vdots \text{ vec } \frac{\partial Y}{\partial x_{21}} \vdots \dots \vdots \text{ vec } \frac{\partial Y}{\partial x_{mn}}\right]' . \tag{5.13}$$

We now use the solution to Example (4.6) where we had established that

$$\text{when } \; Y = X'AX , \quad \text{then } \; \frac{\partial Y}{\partial x_{rs}} = E'_{rs}AX + X'AE_{rs} . \tag{5.14}$$

It follows that

$$\begin{aligned} \text{vec } \frac{\partial Y}{\partial x_{rs}} &= \text{vec } E'_{rs}AX + \text{vec } \; X'AE_{rs} \\ &= (X'A' \otimes I) \; \text{vec } E'_{rs} + (I \otimes X'A) \text{ vec } E_{rs} \end{aligned} \tag{5.15}$$

(using (2.13)) .

Substituting (5.15) into (5.13) we obtain

$$\begin{aligned} \frac{\partial \text{ vec } Y}{\partial \text{ vec } X} &= [(X'A' \otimes I)[\text{vec } E'_{11} \vdots \text{ vec } E'_{21} \vdots \dots \vdots \text{ vec } E'_{mn}]]' \\ &+ [(I \otimes X'A)[\text{vec } E_{11} \vdots \text{ vec } E_{21} \vdots \dots \vdots \text{ vec } E_{mn}]]' \\ &= [\text{vec } E'_{11} \vdots \text{ vec } E'_{21} \vdots \dots \vdots \text{ vec } E'_{mn}]'(AX \otimes I) \\ &+ [\text{vec } E_{11} \vdots \text{ vec } E_{21} \vdots \text{ vec } E_{mn}]'(I \otimes A'X) \end{aligned} \tag{5.16}$$

(by (2.10)).

The matrix

$$[\text{vec } E_{11} \vdots \text{ vec } E_{21} \; \dots \; \vdots \text{ vec } E_{mn}]$$

is the unit matrix I of order $(mn \times mn)$.

Using (2.23) we can write (5.16) as

$$\frac{\partial \text{ vec } Y}{\partial \text{ vec } X} = U'(AX \otimes I) + (I \otimes A'X) .$$

That is

$$\frac{\partial \text{ vec } Y}{\partial \text{ vec } X} = (AX \otimes I)_{(n)} + (I \otimes A'X) . \tag{5.17}$$

In the above calculations we have used the derivative $\partial Y/\partial x_{rs}$ to obtain $(\partial \text{ vec } Y)/(\partial \text{ vec } X)$.

The Second Transformation Principle

Only slight modifications are needed to generalise the above calculations and show that whenever

$$\frac{\partial Y}{\partial x_{rs}} = AE_{rs}B + CE'_{rs}D$$

where A, B, C and D may be functions of X, then

$$\frac{\partial \operatorname{vec} Y}{\partial \operatorname{vec} X} = B \otimes A' + (D \otimes C')_{(n)} \tag{5.18}$$

We will refer to the above result as the **second transformation principle.**

Example 5.2

Find

$$\frac{\partial \operatorname{vec} Y}{\partial \operatorname{vec} X}$$

when

(i) $Y = X'X$ (ii) $Y = AX^{-1}B$.

Solution

Let $\mathbf{y} = \operatorname{vec} Y$ and $\mathbf{x} = \operatorname{vec} X$.

(i) From Example 4.8

$$\frac{\partial Y}{\partial x_{rs}} = E'_{rs}X + X'E_{rs}$$

Now use the second transformation principle, to obtain

$$\frac{\partial \mathbf{y}}{\partial \mathbf{x}} = I \otimes X + (X \otimes I)_{(n)} \quad .$$

(ii) From Example 4.6

$$\frac{\partial Y}{\partial x_{rs}} = -AX^{-1}E_{rs}X^{-1}B$$

hence

$$\frac{\partial \mathbf{y}}{\partial \mathbf{x}} = -(X^{-1}B) \otimes (X^{-1})'A'.$$

Hopefully, using the above results for matrices, we should be able to rediscover results for the derivatives of vectors considered in Chapter 4.

For example let X be a column vector $\mathbf{x}$ then

$$Y = X'X \quad \text{becomes} \quad y = \mathbf{x}'\mathbf{x} \quad (y \text{ is a scalar})\,.$$

The above result for $\partial y/\partial \mathbf{x}$ becomes

$$\frac{\partial y}{\partial \mathbf{x}} = (I \otimes \mathbf{x}) + (\mathbf{x} \otimes I)_{(1)}\,.$$

But the unit vectors involved are of order $(n \times 1)$ which, for the one column vector X is (1×1). Hence

$$\frac{\partial y}{\partial \mathbf{x}} = 1 \otimes \mathbf{x} + \mathbf{x} \otimes 1 \qquad (\text{use } (5.9))$$

$$= \mathbf{x} + \mathbf{x} = 2\mathbf{x}$$

which is the result found in (4.4).

5.4 MORE ON DERIVATIVES OF SCALAR FUNCTIONS WITH RESPECT TO A MATRIX

In section 4.4 we derived a formula, (4.10), which is useful when evaluating $\partial|Y|/\partial X$ for a large class of scalar matrix functions defined by Y.

Example 5.3

Evaluate the derivatives

$$\text{(i)} \quad \frac{\partial \log |X|}{\partial X} \quad \text{and} \quad \text{(ii)} \quad \frac{\partial |X|^r}{\partial X}\,.$$

Solution

(i) We have

$$\frac{\partial}{\partial x_{rs}}(\log |X|) = \frac{1}{|X|}\frac{\partial |X|}{\partial x_{rs}}\,.$$

From Example 4.4,

$$\frac{\partial |X|}{\partial X} = |X|(X^{-1})' \quad (\text{non-symmetric case})\,.$$

Hence

$$\frac{\partial \log |X|}{\partial X} = (X^{-1})'\,.$$

(ii)

$$\frac{\partial |X|^r}{\partial x_{rs}} = r|X|^{r-1}\frac{\partial |X|}{\partial x_{rs}}\,.$$

Hence

$$\frac{\partial |X|^r}{\partial X} = r|X|^r(X^{-1})' .$$

Traces of matrices form an important class of scalar matrix functions covering a wide range of applications, particularly in statistics in the formulation of least squares and various optimisation problems.

Having discussed the evaluation of the derivative $\partial Y/\partial x_{rs}$ for various products of matrices, we can now apply these results to the evaluation of the derivative

$$\frac{\partial(\text{tr } Y)}{\partial X} .$$

We first note that

$$\frac{\partial(\text{tr } Y)}{\partial X} = \left[\frac{\partial(\text{tr } Y)}{\partial x_{rs}}\right] \tag{5.19}$$

where the bracket on the right hand side of (5.19) denotes, (as usual) a matrix of the same order as X, defined by its (r, s)th element.

As a consequence of (5.19) or perhaps more clearly seen from the definition (4.7), we note that on transposing X, we have

$$\frac{\partial(\text{tr } Y)}{\partial X'} = \left(\frac{\partial(\text{tr } Y)}{\partial X}\right)' . \tag{5.20}$$

Another, and possibly an obvious property of a trace is found when considering the definition of $\partial Y/\partial x_{rs}$ (see (4.19)).

Assuming that $Y = [y_{ij}]$ is of order $(n \times n)$

$$\text{tr } \frac{\partial Y}{\partial x_{rs}} \equiv \frac{\partial y_{11}}{\partial x_{rs}} + \frac{\partial y_{22}}{\partial x_{rs}} + \ldots + \frac{\partial y_{nn}}{\partial x_{rs}}$$

$$= \frac{\partial}{\partial x_{rs}}(y_{11} + y_{22} + \ldots + y_{nn}) .$$

Hence,

$$\text{tr } \frac{\partial Y}{\partial x_{rs}} = \frac{\partial(\text{tr } Y)}{\partial x_{rs}} . \tag{5.21}$$

Example 5.4

Evaluate

$$\frac{\partial \text{ tr}(AX)}{\partial X} .$$

Solution

$$\frac{\partial \operatorname{tr}(AX)}{\partial x_{rs}} = \operatorname{tr} \frac{\partial (AX)}{\partial x_{rs}} \qquad \text{by (5.21)}$$

$$= \operatorname{tr}(AE_{rs}) \qquad \text{by Example (4.8)}$$

$$= \operatorname{tr}(E'_{rs}A') \qquad \text{since } \operatorname{tr} Y = \operatorname{tr} Y'$$

$$= (\operatorname{vec} E_{rs})' (\operatorname{vec} A') \text{ by Example (1.4).}$$

Hence,

$$\frac{\partial \operatorname{tr}(AX)}{\partial X} = A' \ .$$

As we found in the previous chapter we can use the derivative of the trace of one product to obtain the derivative of the trace of a different product.

Example 5.5

Evaluate

$$\frac{\partial \operatorname{tr}(AX')}{\partial X} \ .$$

Solution

From the previous result

$$\frac{\partial \operatorname{tr}(BX)}{\partial X} = \frac{\partial \operatorname{tr}(X'B')}{\partial X} = B' \ .$$

Let $A' = B$ in the above equation, it follows that

$$\frac{\partial \operatorname{tr}(X'A)}{\partial X} = \frac{\partial \operatorname{tr}(A'X)}{\partial X} = A.$$

The derivatives of traces of more complicated matrix products can be found similarly.

Example 5.6

Evaluate

$$\frac{\partial (\operatorname{tr} Y)}{\partial X} ,$$

when

(i) $Y = X'AX$
(ii) $Y = X'AXB$

Solution

It is obvious that (i) follows from (ii) when $B = I$.

(ii) $Y = X_1 B$ where $X_1 = X'AX$.

$$\frac{\partial Y}{\partial x_{rs}} = \frac{\partial X_1}{\partial x_{rs}} B$$

$$= E'_{rs} AXB + X'AE_{rs}B \qquad \text{(by Example 4.6)}$$

Hence,

$$\text{tr}\left(\frac{\partial Y}{\partial x_{rs}}\right) = \text{tr}\,(E'_{rs}AXB) + \text{tr}\,(X'AE_{rs}B)$$

$$= \text{tr}\,(E'_{rs}AXB) + \text{tr}\,(E'_{rs}A'XB')$$

$$= (\text{vec}\,E_{rs})'\,\text{vec}\,(AXB) + (\text{vec}\,E_{rs})'\,\text{vec}\,(A'XB')\,.$$

It follows that

$$\frac{\partial(\text{tr}\,Y)}{\partial X} = AXB + A'XB'\,.$$

(i) Let $B = I$ in the above equation, we obtain

$$\frac{\partial(\text{tr}\,Y)}{\partial X} = AX + A'X = (A + A')X\,.$$

5.5 THE MATRIX DIFFERENTIAL

For a scalar function $f(\mathbf{x})$ where $\mathbf{x} = [x_1\, x_2 \ldots x_n]'$, the differential $\text{d}f$ is defined as

$$\text{d}f = \sum_{i=1}^{n} \frac{\partial f}{\partial x_i} \text{d}x_i\,. \tag{5.23}$$

Corresponding to this definition we define the **matrix differential** $\text{d}X$ for the matrix $X = [x_{ij}]$ of order $(m \times n)$ to be

$$\text{d}X = \begin{bmatrix} \text{d}x_{11} & \text{d}x_{12} & \ldots & \text{d}x_{1n} \\ \text{d}x_{21} & \text{d}x_{22} & \ldots & \text{d}x_{2n} \\ \vdots & & & \\ \text{d}x_{m1} & \text{d}x_{m2} & \ldots & \text{d}x_{mn} \end{bmatrix}. \tag{5.24}$$

The following two results follow immediately:

$$\text{d}(\alpha X) = \alpha(\text{d}X) \qquad \text{(where } \alpha \text{ is a scalar)} \tag{5.25}$$

$$\text{d}(X + Y) = \text{d}X + \text{d}Y\,. \tag{5.26}$$

Consider now $X = [x_{ij}]$ of order $(m \times n)$ and $Y = [y_{ij}]$ of order $(n \times p)$.

$$XY = [\sum_j x_{ij}y_{jk}]\,,$$

hence

$$\begin{aligned}\mathrm{d}(XY) &= \mathrm{d}[\sum_j x_{ij}y_{jk}] \\ &= [\sum_j \mathrm{d}(x_{ij}y_{jk})] \\ &= [\sum_j(\mathrm{d}x_{ij})y_{jk}] + [\sum_j x_{ij}(\mathrm{d}y_{jk})] \ .\end{aligned}$$

It follows that

$$\mathrm{d}(XY) = (\mathrm{d}X)Y + X(\mathrm{d}Y) \ . \tag{5.27}$$

Example 5.7

Given $X = [x_{ij}]$ a nonsingular matrix, evaluate

(i) $\mathrm{d}|X|$, (ii) $\mathrm{d}(X^{-1})$

Solution

(i) By (5.23)

$$\begin{aligned}\mathrm{d}|X| &= \sum_{i,j} \frac{\partial |X|}{\partial x_{ij}} (\mathrm{d}x_{ij}) \\ &= \sum_{i,j} X_{ij}(\mathrm{d}x_{ij})\end{aligned}$$

since $(\partial |X|)/(\partial x_{ij}) = X_{ij}$, the cofactor of x_{ij} in $|X|$.

By an argument similar to the one used in section 4.4, we can write

$$\mathrm{d}|X| = \mathrm{tr}\ \{Z'(\mathrm{d}X)\} \qquad \text{(compare with (4.10))}$$

where $Z = [X_{ij}]$.

Since $Z' = |X|X^{-1}$, we can write

$$\mathrm{d}|X| = |X|\ \mathrm{tr}\ \{X^{-1}(\mathrm{d}X)\} \ .$$

(ii) Since

$$X^{-1}X = I$$

we use (5.27) to write

$$\mathrm{d}(X^{-1})X + X^{-1}(\mathrm{d}X) = 0 \ .$$

Hence

$$\mathrm{d}(X^{-1}) = -X^{-1}(\mathrm{d}X)X^{-1}$$

(compare with Example 4.6).

Notice that if X is a **symmetric** matrix, then

$$X = X'$$

and

$$(\mathrm{d}X)' = \mathrm{d}X \ . \tag{5.28}$$

Problems for Chapter 5

(1) Consider

$$A = \begin{bmatrix} a_{11} & a_{12} \\ a_{21} & a_{22} \end{bmatrix}, \quad X = \begin{bmatrix} x_{11} & x_{12} \\ x_{21} & x_{22} \end{bmatrix} \text{ and } Y = AX'.$$

Use a direct method to evaluate

$$\frac{\partial \operatorname{vec} Y}{\partial \operatorname{vec} X}$$

and verify (5.10).

(2) Obtain

$$\frac{\partial \operatorname{vec} Y}{\partial \operatorname{vec} X},$$

when

(i) $Y = AX'B$ and (ii) $Y = X^2$.

(3) Find expressions for

$$\frac{\partial \operatorname{tr} Y}{\partial X},$$

when

(a) $Y = AXB$, (b) $Y = X^2$ and (c) $Y = XX'$.

(4) Evaluate

$$\frac{\partial \operatorname{tr} Y}{\partial X}$$

when

(a) $Y = X^{-1}$, (b) $Y = AX^{-1}B$, (c) $Y = X^n$ and (d) $Y = e^X$.

(5) (a) Use the direct method to obtain expressions for the matrix differential dY when

(i) $Y = AX$, (ii) $Y = X'X$ and (iii) $Y = X^2$.

(b) Find dY when

$$Y = AXBX.$$

CHAPTER 6

The Derivative of a Matrix with respect to a Matrix

6.1 INTRODUCTION

In the previous two chapters we have defined the derivative of a matrix with respect to a scalar and the derivative of a scalar with respect to a matrix. We will now generalise the definitions to include the derivative of a matrix with respect to a matrix. The author has adopted the definition suggested by Vetter [31], although other definitions also give rise to some useful results.

6.2 THE DEFINITIONS AND SOME RESULTS

Let $Y = [y_{ij}]$ be a matrix of order $(p \times q)$. We have defined (see (4.19)) the derivative of Y with respect to a scalar x_{rs}, it is the matrix $[\partial y_{ij}/\partial x_{rs}]$ of order $(p \times q)$.

Let $X = [x_{rs}]$ be a matrix of order $(m \times n)$ we generalise (4.19) and define the derivative of Y with respect to X, denoted by

$$\frac{\partial Y}{\partial X}$$

as the partitioned matrix whose (r, s)th partition is

$$\frac{\partial Y}{\partial x_{rs}}$$

in other words

$$\frac{\partial Y}{\partial X} = \begin{bmatrix} \dfrac{\partial Y}{\partial x_{11}} & \dfrac{\partial Y}{\partial x_{12}} & \cdots & \dfrac{\partial Y}{\partial x_{1n}} \\ \dfrac{\partial Y}{\partial x_{21}} & \dfrac{\partial Y}{\partial x_{22}} & \cdots & \dfrac{\partial Y}{\partial x_{2n}} \\ \vdots & & & \\ \dfrac{\partial Y}{\partial x_{m1}} & \dfrac{\partial Y}{\partial x_{m2}} & \cdots & \dfrac{\partial Y}{\partial x_{mn}} \end{bmatrix} = \sum_{r,s} E_{rs} \otimes \frac{\partial Y}{\partial x_{rs}} \tag{6.1}$$

The right hand side of (6.1) following from the definitions (1.4) and (2.1) where E_{rs} is of order $(m \times n)$, the order of the matrix X.

It is seen that $\partial Y/\partial X$ is a matrix of order $(mp \times nq)$.

Example 6.1

Consider

$$Y = \begin{bmatrix} x_{11}x_{12}x_{22} & e^{x_{11}x_{22}} \\ \sin(x_{11}+x_{12}) & \log(x_{11}+x_{21}) \end{bmatrix}$$

and

$$X = \begin{bmatrix} x_{11} & x_{12} \\ x_{21} & x_{22} \end{bmatrix},$$

Evaluate

$$\frac{\partial Y}{\partial X}.$$

Solution

$$\frac{\partial Y}{\partial x_{11}} = \begin{bmatrix} x_{12}x_{22} & x_{22}e^{x_{11}x_{22}} \\ \cos(x_{11}+x_{12}) & \dfrac{1}{(x_{11}+x_{21})} \end{bmatrix},$$

$$\frac{\partial Y}{\partial x_{12}} = \begin{bmatrix} x_{11}x_{22} & 0 \\ \cos(x_{11}+x_{12}) & 0 \end{bmatrix},$$

$$\frac{\partial Y}{\partial x_{21}} = \begin{bmatrix} 0 & 0 \\ 0 & \dfrac{1}{x_{11}+x_{21}} \end{bmatrix}, \quad \frac{\partial Y}{\partial x_{22}} = \begin{bmatrix} x_{11}x_{12} & x_{11}e^{x_{11}x_{22}} \\ 0 & 0 \end{bmatrix}.$$

$$\frac{\partial Y}{\partial X} = \begin{bmatrix} x_{12}x_{22} & x_{22}e^{x_{11}x_{22}} & x_{11}x_{22} & 0 \\ \cos(x_{11}+x_{12}) & \dfrac{1}{x_{11}+x_{21}} & \cos(x_{11}+x_{12}) & 0 \\ 0 & 0 & x_{11}x_{12} & x_{11}e^{x_{11}x_{22}} \\ 0 & \dfrac{1}{x_{11}+x_{21}} & 0 & 0 \end{bmatrix}$$

Example 6.2

Given the matrix $X = [x_{ij}]$ of order $(m \times n)$, evaluate $\partial X/\partial X$ when

(i) All elements of X are independent

(ii) X is a symmetric matrix (of course in this case $m = n$).

Solution

(i) By (6.1)

$$\frac{\partial X}{\partial X} = \sum_{r,s} E_{rs} \otimes E_{rs} = \bar{U} \quad \text{(see (2.26))}$$

(ii)

$$\frac{\partial X}{\partial x_{rs}} = E_{rs} + E_{sr} \qquad \text{for} \quad r \neq s$$

$$\frac{\partial X}{\partial x_{rs}} = E_{rr} \qquad \text{for} \quad r = s \quad .$$

We can write the above as;

$$\frac{\partial X}{\partial x_{rs}} = E_{rs} + E_{sr} - \delta_{rs} E_{rr} \, .$$

Hence,

$$\frac{\partial X}{\partial X} = \sum_{r,s} E_{rs} \otimes E_{rs} + \sum_{r,s} E_{rs} \otimes E_{sr} - \delta_{rs} \sum_{r,s} E_{sr} \otimes E_{rr}$$

$$= \bar{U} + U - \Sigma E_{rr} \otimes E_{rr} \qquad \text{(see (2.24) and (2.26))}$$

Example 6.3

Evaluate and write out in full $\partial X'/\partial X$ given

$$X = \begin{bmatrix} x_{11} & x_{12} & x_{13} \\ x_{21} & x_{22} & x_{23} \end{bmatrix} \, .$$

Solution

By (6.1) we have

$$\frac{\partial X'}{\partial X} = E_{rs} \otimes E'_{rs}$$

$$= U \, .$$

Hence

$$\frac{\partial X'}{\partial X} = \begin{bmatrix} 1 & 0 & 0 & 0 & 0 & 0 \\ 0 & 0 & 1 & 0 & 0 & 0 \\ 0 & 0 & 0 & 0 & 1 & 0 \\ 0 & 1 & 0 & 0 & 0 & 0 \\ 0 & 0 & 0 & 1 & 0 & 0 \\ 0 & 0 & 0 & 0 & 0 & 1 \end{bmatrix} \, .$$

From the definition (6.1) we obtain

$$\left(\frac{\partial Y}{\partial X}\right)' = \left(\sum_{r,s} E_{rs} \otimes \frac{\partial Y}{\partial x_{rs}}\right)'$$

$$= \sum_{r,s} E'_{rs} \otimes \left(\frac{\partial Y}{\partial x_{rs}}\right)' \qquad \text{by (2.10)}$$

$$= \sum_{r,s} E'_{rs} \otimes \frac{\partial Y'}{\partial x_{rs}} \qquad \text{from (4.19)}.$$

It follows that

$$\left(\frac{\partial Y}{\partial X}\right)' = \frac{\partial Y'}{\partial X'} . \tag{6.2}$$

6.3 PRODUCT RULES FOR MATRICES

We shall first obtain a rule for the derivative of a product of matrices with respect to a matrix, that is to find an expression for

$$\frac{\partial (XY)}{\partial Z}$$

where the order of the matrices are as indicated

$$X(m \times n)\,, \quad Y(n \times v)\,, \quad Z(p \times q)\,.$$

By (4.24) we write

$$\frac{\partial (XY)}{\partial z_{rs}} = \frac{\partial X}{\partial z_{rs}} Y + X \frac{\partial Y}{\partial z_{rs}}$$

where $Z = [z_{rs}]$.

If E_{rs} is an elementary matrix of order $(p \times q)$, we make use of (6.1) to write

$$\frac{\partial (XY)}{\partial Z} = \sum_{r,s} E_{rs} \otimes \left[\frac{\partial X}{\partial z_{rs}} Y + X \frac{\partial Y}{\partial z_{rs}}\right]$$

$$= \sum_{r,s} E_{rs} \otimes \frac{\partial X}{\partial z_{rs}} Y + \sum_{r,s} E_{rs} \otimes X \frac{\partial Y}{\partial z_{rs}}$$

$$= \sum_{r,s} E_{rs} I_q \otimes \frac{\partial X}{\partial z_{rs}} Y + \sum_{r,s} I_p E_{rs} \otimes X \frac{\partial Y}{\partial z_{rs}}$$

(where I_q and I_p are unit matrices of order $(q \times q)$ and $(p \times p)$ respectively)

$$= \sum_{r,s} \left(E_{rs} \otimes \frac{\partial X}{\partial z_{rs}}\right)(I_q \otimes Y) + \sum_{r,s} (I_p \otimes X)\left(E_{rs} \otimes \frac{\partial Y}{\partial z_{rs}}\right) \quad \text{(by 2.11)}$$

finally, by (6.1)

$$\frac{\partial(XY)}{\partial Z} = \frac{\partial X}{\partial Z}(I_q \otimes Y) + (I_p \otimes X)\frac{\partial Y}{\partial Z} \tag{6.3}$$

Example 6.4

Find an expression for

$$\frac{\partial X^{-1}}{\partial X}.$$

Solution

Using (6.3) on

$$XX^{-1} = I,$$

we obtain

$$\frac{\partial(XX^{-1})}{\partial X} = \frac{\partial X}{\partial X}(I \otimes X^{-1}) + (I \otimes X)\frac{\partial X^{-1}}{\partial X} = 0$$

hence

$$\frac{\partial X^{-1}}{\partial X} = -(I \otimes X)^{-1}\frac{\partial X}{\partial X}(I \otimes X^{-1})$$

$$= -(I \otimes X^{-1})\,\bar{U}\,(I \otimes X^{-1})$$

(by Example 6.2 and (2.12)).

Next we determine a rule for the derivative of a Kronecker product of matrices with respect to a matrix, that is an expression for

$$\frac{\partial(X \otimes Y)}{\partial Z}$$

The order of the matrix Y is not now restricted, we will consider that it is $(u \times v)$. On representing $X \otimes Y$ by the (i,j)th partition $[x_{ij}Y]$ $(i = 1, 2, \ldots, m,$ $j = 1, 2, \ldots, n)$, we can write

$$\frac{\partial(X \otimes Y)}{\partial z_{rs}} = \frac{\partial}{\partial z_{rs}}[x_{ij}Y]$$

where (r, s) are fixed

$$= \left[\frac{\partial x_{ij}}{\partial z_{rs}} Y\right] + \left[x_{ij} \frac{\partial Y}{\partial z_{rs}}\right]$$

$$= \frac{\partial X}{\partial z_{rs}} \otimes Y + X \otimes \frac{\partial Y}{\partial z_{rs}} .$$

Hence by (6.1)

$$\frac{\partial (X \otimes Y)}{\partial Z} = \sum_{r,s} E_{rs} \otimes \frac{\partial X}{\partial z_{rs}} \otimes Y + \sum_{r,s} E_{rs} \otimes X \otimes \frac{\partial Y}{\partial z_{rs}}$$

where E_{rs} is of order $(p \times q)$

$$= \frac{\partial X}{\partial Z} \otimes Y + \sum_{r,s} E_{rs} \otimes \left(X \otimes \frac{\partial Y}{\partial z_{rs}}\right) .$$

The summation on the right hand side is not $X \otimes \partial Y/\partial Z$ as may appear at first sight, nevertheless it can be put into a more convenient form, as a product of matrices. To achieve this aim we make repeated use of (2.8) and (2.11)

$$\sum_{r,s} E_{rs} \otimes \left(X \otimes \frac{\partial Y}{\partial z_{rs}}\right) = \sum_{r,s} [I_p E_{rs} I_q] \otimes \left[U_1\left(\frac{\partial Y}{\partial z_{rs}} \otimes X\right) U_2\right] \quad \text{by (2.14)}$$

$$= \sum_{r,s} \left[(I_p E_{rs}) \otimes U_1\left(\frac{\partial Y}{\partial z_{rs}} \otimes X\right)\right] [I_q \otimes U_2] \quad \text{by (2.11)}$$

$$= \sum_{r,s} [I_p \otimes U_1] \left\{E_{rs} \otimes \frac{\partial Y}{\partial z_{rs}} \otimes X\right\} [I_q \otimes U_2] \text{ by (2.11)} .$$

Hence

$$\frac{\partial (X \otimes Y)}{\partial Z} = \frac{\partial X}{\partial Z} \otimes Y + [I_p \otimes U_1] \left[\frac{\partial Y}{\partial Z} \otimes X\right] [I_q \otimes U_2] \tag{6.4}$$

where U_1 and U_2 are permutation matrices of orders $(mu \times mu)$ and $(nv \times nv)$ repectively.

We illustrate the use of equation (6.4) with a simple example.

Example 6.5

$A = [a_{ij}]$ and $X = [x_{ij}]$ are matrices, each of order (2×2). Use

(i) Equation (6.4), and
(ii) a direct method to evaluate

$$\frac{\partial (A \otimes X)}{\partial X} .$$

Solution

(i) In this example (6.4) becomes

$$\frac{\partial(A \otimes X)}{\partial X} = [I \otimes U_1]\left[\frac{\partial X}{\partial X} \otimes A\right][I \otimes U_2]$$

where I is the unit matrix of order (2×2) and

$$U_1 = U_2 = \Sigma E_{rs} \otimes E'_{rs} = \begin{bmatrix} 1 & 0 & 0 & 0 \\ 0 & 0 & 1 & 0 \\ 0 & 1 & 0 & 0 \\ 0 & 0 & 0 & 1 \end{bmatrix}.$$

Since

$$\frac{\partial X}{\partial X} = \bar{U} = \begin{bmatrix} 1 & 0 & 0 & 1 \\ 0 & 0 & 0 & 0 \\ 0 & 0 & 0 & 0 \\ 1 & 0 & 0 & 1 \end{bmatrix}$$

only a simple calculation is necessary to obtain the result. It is found that

$$\frac{\partial(A \otimes X)}{\partial X} = \begin{bmatrix} a_{11} & 0 & a_{12} & 0 & 0 & a_{11} & 0 & a_{12} \\ 0 & 0 & 0 & 0 & 0 & 0 & 0 & 0 \\ a_{21} & 0 & a_{22} & 0 & 0 & a_{21} & 0 & a_{22} \\ 0 & 0 & 0 & 0 & 0 & 0 & 0 & 0 \\ 0 & 0 & 0 & 0 & 0 & 0 & 0 & 0 \\ a_{11} & 0 & a_{12} & 0 & 0 & a_{11} & 0 & a_{12} \\ 0 & 0 & 0 & 0 & 0 & 0 & 0 & 0 \\ a_{21} & 0 & a_{22} & 0 & 0 & a_{21} & 0 & a_{22} \end{bmatrix}.$$

(ii) We evaluate

$$Y = A \otimes X = \begin{bmatrix} a_{11}x_{11} & a_{11}x_{12} & a_{12}x_{11} & a_{12}x_{12} \\ a_{11}x_{21} & a_{11}x_{22} & a_{12}x_{21} & a_{12}x_{22} \\ a_{21}x_{11} & a_{21}x_{12} & a_{22}x_{11} & a_{22}x_{12} \\ a_{21}x_{21} & a_{21}x_{22} & a_{22}x_{21} & a_{22}x_{22} \end{bmatrix}$$

and then make use of (6.1) to obtain the above result.

6.4 THE CHAIN RULE FOR THE DERIVATIVE OF A MATRIX WITH RESPECT TO A MATRIX

We wish to obtain an expression for

$$\frac{\partial Z}{\partial X}$$

where the matrix Z is a matrix function of a matrix X, that is

$$Z = Z(Y(X))$$

where

$$X = [x_{ij}] \text{ is of order } (m \times n)$$
$$Y = [y_{ij}] \text{ is of order } (u \times v)$$
$$Z = [z_{ij}] \text{ is of order } (p \times q)$$

By definition in (6.1)

$$\frac{\partial Z}{\partial X} = \sum_{r,s} E_{rs} \otimes \frac{\partial Z}{\partial x_{rs}} \qquad \begin{array}{l} r = 1, 2, \ldots, m \\ s = 1, 2, \ldots, n \end{array} .$$

where E_{rs} is an elementary matrix of order $(m \times n)$,

$$= \sum_{r,s} E_{rs} \otimes \sum_{i,j} E_{ij} \frac{\partial z_{ij}}{\partial x_{rs}} \qquad \begin{array}{l} i = 1, 2, \ldots, p \\ j = 1, 2, \ldots, q \end{array}$$

where E_{ij} is of order $(p \times q)$

As in section 4.3, we use the chain rule to write

$$\frac{\partial z_{ij}}{\partial x_{rs}} = \sum_{\alpha,\beta} \frac{\partial z_{ij}}{\partial y_{\alpha\beta}} \cdot \frac{\partial y_{\alpha\beta}}{\partial x_{rs}} \qquad \begin{array}{l} \alpha = 1, 2, \ldots, u \\ \beta = 1, 2, \ldots, v \end{array}$$

Hence

$$\frac{\partial Z}{\partial X} = \sum_{r,s} E_{rs} \otimes \left[\sum_{i,j} E_{ij} \sum_{\alpha,\beta} \frac{\partial z_{ij}}{\partial y_{\alpha\beta}} \cdot \frac{\partial y_{\alpha\beta}}{\partial x_{rs}} \right]$$

$$= \sum_{r,s} \sum_{\alpha,\beta} E_{rs} \frac{\partial y_{\alpha\beta}}{\partial x_{rs}} \otimes \sum_{i,j} E_{ij} \frac{\partial z_{ij}}{\partial y_{\alpha\beta}} \qquad \text{(by 2.5)}$$

$$= \sum_{\alpha,\beta} \frac{\partial y_{\alpha\beta}}{\partial X} \otimes \frac{\partial Z}{\partial y_{\alpha\beta}} \qquad \text{(by (4.7) and (4.19))} .$$

If I_n and I_p are unit matrices of orders $(n \times n)$ and $(p \times p)$ respectively, we can write the above as

$$\frac{\partial Z}{\partial X} = \sum_{\alpha,\beta}\left(\frac{\partial y_{\alpha\beta}}{\partial X} I_n\right) \otimes \left(I_p \frac{\partial Z}{\partial y_{\alpha\beta}}\right).$$

Hence, by (2.11)

$$\frac{\partial Z}{\partial X} = \sum_{\alpha,\beta}\left(\frac{\partial y_{\alpha\beta}}{\partial X} \otimes I_p\right)\left(I_n \otimes \frac{\partial Z}{\partial y_{\alpha\beta}}\right) \tag{6.5}$$

Equation (6.5) can be written in a more convenient form, avoiding the summation, if we define an appropriate notation, a generalisation of the previous one.

Since

$$Y = \begin{bmatrix} y_{11} & y_{12} & \dots & y_{1v} \\ y_{21} & y_{22} & \dots & y_{2v} \\ & & & \\ y_{u1} & y_{u2} & \dots & y_{uv} \end{bmatrix}$$

than $(\text{vec } Y)' = [y_{11}\; y_{21} \dots y_{uv}]$.

We will write the partitioned matrix

$$\left[\frac{\partial y_{11}}{\partial X} \otimes I_p \;\vdots\; \frac{\partial y_{21}}{\partial X} \otimes I_p \;\vdots\; \dots \;\frac{\partial y_{uv}}{\partial X} \otimes I_p\right]$$

as

$$\frac{\partial [y_{11}\; y_{21} \dots y_{uv}]}{\partial X} \otimes I_p$$

or as

$$\frac{\partial (\text{vec } Y)'}{\partial X} \otimes I_p \;.$$

Similarly, we write the partitioned matrix

$$\begin{bmatrix} I_n \otimes \dfrac{\partial Z}{\partial y_{11}} \\ \hline I_n \otimes \dfrac{\partial Z}{\partial y_{21}} \\ \hline \vdots \\ \hline I_n \otimes \dfrac{\partial Z}{\partial y_{uv}} \end{bmatrix} \quad \text{as} \quad \left[I_n \otimes \frac{\partial Z}{\partial \text{ vec } Y}\right]$$

We can write the sum (6.5) in the following order

$$\frac{\partial Z}{\partial X} = \left[\frac{\partial y_{11}}{\partial X} \otimes I_p\right]\left[I_n \otimes \frac{\partial Z}{\partial y_{11}}\right] + \left[\frac{\partial y_{21}}{\partial X} \otimes I_p\right]\left[I_n \otimes \frac{\partial Z}{y_{21}}\right] + \dots$$

$$+ \left[\frac{\partial y_{uv}}{\partial X} \otimes I_p\right]\left[I_n \otimes \frac{\partial Z}{\partial y_{uv}}\right].$$

We can write this as a (partitioned) matrix product

$$\frac{\partial Z}{\partial X} = \left[\frac{\partial y_{11}}{\partial X} \otimes I_p \vdots \frac{\partial y_{21}}{\partial X} \otimes I_p \vdots \dots \vdots \frac{\partial y_{uv}}{\partial X} \otimes I_p\right] \begin{bmatrix} I_n \otimes \dfrac{\partial Z}{\partial y_{11}} \\ \hdashline I_n \otimes \dfrac{\partial Z}{\partial y_{21}} \\ \hdashline \vdots \\ \hdashline I_n \otimes \dfrac{\partial Z}{\partial y_{uv}} \end{bmatrix}.$$

Finally, using the notations defined above, we have

$$\frac{\partial Z}{\partial X} = \left[\frac{\partial\,[\text{vec}\, Y]'}{\partial X} \otimes I_p\right]\left[I_n \otimes \frac{\partial Z}{\partial\, \text{vec}\, Y}\right] \tag{6.6}$$

We consider a simple example to illustrate the application of the above formula. The example can also be solved by evaluating the matrix Z in terms of the components of the matrix X and then applying the definition in (6.1).

Example 6.6

Given the matrix $A = [a_{ij}]$ and $X = [x_{ij}]$ both of order (2×2), evaluate

$$\partial Z/\partial X$$

where $Z = Y'Y$ and $Y = AX$.

(i) Using (6.6)

(ii) Using a direct method.

Solution

(i) For convenience write (6.6) as

$$\frac{\partial Z}{\partial X} = QR$$

where

$$Q = \left[\frac{\partial\,[\text{vec}\, Y]'}{\partial X} \otimes I_p\right] \quad \text{and} \quad R = \left[I_n \otimes \frac{\partial Z}{\partial\, \text{vec}\, Y}\right].$$

From Example 4.8 we know that

$$\frac{\partial y_{ij}}{\partial X} = A'E_{ij}$$

so that Q can now be easily evaluated,

$$Q = \left[\begin{array}{cccc|cccc|cccc|cccc} a_{11} & 0 & 0 & 0 & a_{21} & 0 & 0 & 0 & 0 & 0 & a_{11} & 0 & 0 & 0 & a_{21} & 0 \\ 0 & a_{11} & 0 & 0 & 0 & a_{21} & 0 & 0 & 0 & 0 & 0 & a_{11} & 0 & 0 & 0 & a_{21} \\ a_{12} & 0 & 0 & 0 & a_{22} & 0 & 0 & 0 & 0 & 0 & a_{12} & 0 & 0 & 0 & a_{22} & 0 \\ 0 & a_{12} & 0 & 0 & 0 & a_{22} & 0 & 0 & 0 & 0 & 0 & a_{12} & 0 & 0 & 0 & a_{22} \end{array}\right].$$

Also in Example 4.8 we have found

$$\frac{\partial Z}{\partial y_{rs}} = E'_{rs}Y + Y'E_{rs}$$

we can now evaluate R

$$R = \left[\begin{array}{cccc} 2y_{11} & y_{12} & 0 & 0 \\ y_{12} & 0 & 0 & 0 \\ 0 & 0 & 2y_{11} & y_{12} \\ 0 & 0 & y_{12} & 0 \\ \hline 2y_{21} & y_{22} & 0 & 0 \\ y_{22} & 0 & 0 & 0 \\ 0 & 0 & 2y_{21} & y_{22} \\ 0 & 0 & y_{22} & 0 \\ \hline 0 & y_{11} & 0 & 0 \\ y_{11} & 2y_{12} & 0 & 0 \\ 0 & 0 & 0 & y_{11} \\ 0 & 0 & y_{11} & 2y_{12} \\ \hline 0 & y_{21} & 0 & 0 \\ y_{21} & 2y_{22} & 0 & 0 \\ 0 & 0 & 0 & y_{21} \\ 0 & 0 & y_{21} & 2y_{22} \end{array}\right].$$

The product of Q and R is the derivative we have been asked to evaluate

$$QR = \begin{bmatrix} 2a_{11}y_{11}+2a_{21}y_{21} & a_{11}y_{12}+a_{21}y_{22} & 0 & a_{11}y_{11}+a_{21}y_{21} \\ a_{11}y_{12}+a_{21}y_{22} & 0 & a_{11}y_{11}+a_{21}y_{21} & 2a_{11}y_{12}+2a_{21}y_{22} \\ 2a_{12}y_{11}+2a_{22}y_{21} & a_{12}y_{12}+a_{22}y_{22} & 0 & a_{12}y_{11}+a_{22}y_{21} \\ a_{12}y_{12}+a_{22}y_{22} & 0 & a_{12}y_{11}+a_{22}y_{21} & 2a_{12}y_{12}+2a_{22}y_{22} \end{bmatrix}$$

(ii) By a simple extension of the result of Example 4.6(ii) we find that when

$$Z = X'A'AX$$

$$\frac{\partial Z}{\partial x_{rs}} = E'_{rs}A'AX + X'A'AE_{rs}$$

$$= E'_{rs}A'Y + Y'AE_{rs}$$

where $Y = AX$.

By (6.1) and (2.11)

$$\frac{\partial Z}{\partial X} = \sum_{r,s}(E_{rs}\otimes E'_{rs})(I\otimes A'Y) + \sum_{r,s}(I\otimes Y'A)(E_{rs}\otimes E_{rs}) .$$

Since the matrices involved are all of order (2×2)

$$\Sigma E_{rs}\otimes E'_{rs} = \begin{bmatrix} 1 & 0 & 0 & 0 \\ 0 & 0 & 1 & 0 \\ 0 & 1 & 0 & 0 \\ 0 & 0 & 0 & 1 \end{bmatrix}$$

and

$$\Sigma E_{rs}\otimes E_{rs} = \begin{bmatrix} 1 & 0 & 0 & 1 \\ 0 & 0 & 0 & 0 \\ 0 & 0 & 0 & 0 \\ 1 & 0 & 0 & 1 \end{bmatrix} .$$

On substitution and multiplying out in the above expression for $\partial Z/\partial X$, we obtain the same matrix as in (i).

Problems for Chapter 6

(1) Evaluate $\partial Y/\partial X$ given

$$Y = \begin{bmatrix} \cos(x_{12}+x_{22}) & x_{11}x_{21} \\ e^{x_{11}x_{12}} & x_{12}x_{22} \end{bmatrix} \quad \text{and} \quad X = \begin{bmatrix} x_{11} & x_{12} \\ x_{21} & x_{22} \end{bmatrix}$$

(2) The elements of the matrix $X = \begin{bmatrix} x_{11} & x_{21} \\ x_{12} & x_{22} \\ x_{13} & x_{23} \end{bmatrix}$

are all independent. Use a direct method to evaluate $\partial X/\partial X$.

(3) Given a non-singular matrix $X = \begin{bmatrix} x_{11} & x_{12} \\ x_{21} & x_{22} \end{bmatrix}$

use a direct method to obtain

$$\frac{\partial X^{-1}}{\partial X}$$

and verify the solution to Example 6.4.

(4) The matrices $A = [a_{ij}]$ and $X = [x_{ij}]$ are both of order (2×2), X is non-singular. Use a direct method to evaluate

$$\frac{\partial (A \otimes X^{-1})}{\partial X} .$$

CHAPTER 7

Some Applications of Matrix Calculus

7.1 INTRODUCTION

As in Chapter 3, where a number of applications of the Kronecker product were considered, in this chapter a number of applications of matrix calculus are discussed. The applications have been selected from a number considered in the published literature, as indicated in the Bibliography at the end of this book.

These problems were originally intended for the expert, but by expansion and simplification it is hoped that they will now be appreciated by the general reader.

7.2 THE PROBLEMS OF LEAST SQUARES AND CONSTRAINED OPTIMISATION IN SCALAR VARIABLES

In this section we consider, very briefly, the Method of Least Squares to obtain a curve or a line of 'best fit', and the Method of Lagrange Multipliers to obtain an extremum of a function subject to constraints.

For the least squares method we consider a set of data

$$(x_i, y_i) \quad i = 1, 2, \ldots, n \tag{7.1}$$

and a relationship, usually a polynomial function

$$y = f(x)\,. \tag{7.2}$$

For each x_i, we evaluate $f(x_i)$ and the **residual** or the **deviation**

$$e_i = y_i - f(x_i)\,. \tag{7.3}$$

The method depends on choosing the unknown parameters, the polynomial coefficients when $f(x)$ is a polynomial, so that the sum of the squares of the residuals is a minimum, that is

$$S = \sum_{i=1}^{n} e_i^2 = \sum_{i=1}^{n} (y_i - f(x_i))^2 \tag{7.4}$$

is a minimum.

In particular, when $f(x)$ is a linear function

$$y = a_0 + a_1 x$$

$S(a_0, a_1)$ is a minimum when

$$\frac{\partial S}{\partial a_0} = 0 = \frac{\partial S}{\partial a_1}\ . \tag{7.5}$$

These two equations, known as **normal** equations, determine the two unknown parameters a_0 and a_1 which specify the line of 'best fit' according to the principle of least squares.

For the second method we wish to determine the extremum of a continuously differentiable function

$$f(x_1, x_2, \ldots, x_n) \tag{7.6}$$

whose n variables are contrained by m equations of the form

$$g_i(x_1, x_2, \ldots, x_n) = 0\ . \qquad i = 1, 2, \ldots, m \tag{7.7}$$

The method of Lagrange Multipliers depends on defining an augmented function

$$f^* = f + \sum_{i=1}^{m} \mu_i g_i \tag{7.8}$$

where the μ_i are known as **Lagrange multipliers.**

The extreme of $f(\mathbf{x})$ is determined by solving the system of the $(m + n)$ equations

$$\begin{aligned} \frac{\partial f^*}{\partial x_r} &= 0 \qquad r = 1, 2, \ldots, n \\ g_i &= 0 \qquad i = 1, 2, \ldots, m \end{aligned} \tag{7.9}$$

for the m parameters $\mu_1, \mu_2, \ldots, \mu_m$ and the n variables $\mathbf{x}$ determining the extremum.

Example 7.1

Given a matrix $A = [a_{ij}]$ of order (2×2) determine a symmetric matrix $X = [x_{ij}]$ which is a best approximation to A by the criterion of least squares.

Solution

Corresponding to (7.3) we have

$$E = A - X$$

where $E = [e_{ij}]$ and $e_{ij} = a_{ij} - x_{ij}$.

The criterion of least squares for this example is to minimise

$$S = \sum_{i,j} e_{ij}^2 = \Sigma(a_{ij} - x_{ij})^2$$

which is the equivalent of (7.6) above.

The constraint equation is

$$x_{12} - x_{21} = 0$$

and the augmented function is

$$f^* = \Sigma(a_{ij} - x_{ij})^2 + \mu(x_{12} - x_{21}) = 0$$

$$\frac{\partial f^*}{\partial x_{11}} = -2(a_{11} - x_{11}) = 0$$

$$\frac{\partial f^*}{\partial x_{12}} = -2(a_{12} - x_{12}) + \mu = 0$$

$$\frac{\partial f^*}{\partial x_{21}} = -2(a_{21} - x_{21}) - \mu = 0$$

$$\frac{\partial f^*}{\partial x_{22}} = -2(a_{22} - x_{22}) = 0$$

This system of 5 equations (including the constraint) leads to the solution

$$\mu = a_{12} - x_{21}$$

$$x_{11} = a_{11}\,, \quad x_{22} = a_{22}\,, \quad x_{12} = x_{21} = \tfrac{1}{2}(a_{12} + a_{21})\,.$$

Hence

$$X = \begin{bmatrix} a_{11} & \dfrac{a_{12}+a_{21}}{2} \\ \dfrac{a_{12}+a_{21}}{2} & a_{22} \end{bmatrix} = \frac{1}{2}\begin{bmatrix} a_{11} & a_{12} \\ a_{21} & a_{22} \end{bmatrix} + \frac{1}{2}\begin{bmatrix} a_{11} & a_{21} \\ a_{12} & a_{22} \end{bmatrix}$$

$$= \tfrac{1}{2}(A + A')$$

7.3 PROBLEM 1 – MATRIX CALCULUS APPROACH TO THE PROBLEMS OF LEAST SQUARES AND CONSTRAINED OPTIMISATION

If we can express the residuals in the form of a matrix E, as in Example 7.1, then the sum of the residuals squared is

$$S = \operatorname{tr} E'E \quad . \tag{7.10}$$

The criterion of the least squares method is to minimise (7.10) with respect to the parameters involved.

The constrained optimisation problem then takes the form of finding the matrix X such that the scalar matrix function

$$S = f(X)$$

is minimised subject to contraints on X in the form of

$$G(X) = 0 \tag{7.11}$$

where $G = [g_{ij}]$ is a matrix of order $(s \times t)$ where s and t are dependent on the number of constraints g_{ij} involved.

As for the scalar case, we use Lagrange multipliers to form an augmented matrix function $f^*(X)$.

Each constraint g_{ij} is associated with a parameter (Lagrange multiplier) μ_{ij}.

Since

$$\Sigma \mu_{ij} g_{ij} = \operatorname{tr} U'G$$

where

$$U = [\mu_{ij}]$$

we can write the augmented scalar matrix function as

$$f^*(X) = \operatorname{tr} E'E + \operatorname{tr} U'G \tag{7.12}$$

which is the equivalent to (7.8). To find the optimal X, we must solve the system of equations

$$\frac{\partial f^*}{\partial X} = 0\,. \tag{7.13}$$

Problem

Given a non-singular matrix $A = [a_{ij}]$ of order $(n \times n)$ determine a matrix $X = [x_{ij}]$ which is a least squares approximation to A

(i) when X is a symmetric matrix
(ii) when X is an orthogonal matrix.

Solution

(i) The problem was solved in Example 7.1 when A and X are of order (2×2). With the terminology defined above, we write

$$E = A - X$$

$$G(X) = X - X' = 0$$

so that G and hence U are both of order $(n \times n)$.

Equation (7.12) becomes

$$f^* = \operatorname{tr}[A'-X'][A-X] + \operatorname{tr} U'[X-X']$$
$$= \operatorname{tr} A'A - \operatorname{tr} A'X - \operatorname{tr} X'A + \operatorname{tr} X'X + \operatorname{tr} U'X - \operatorname{tr} U'X'.$$

We now make use of the results, in modified form if necessary, of Examples 5.4 and 5.5, we obtain

$$\frac{\partial f^*}{\partial X} = -2A + 2X + U - U'$$
$$= 0 \quad \text{for} \quad X = A + \frac{U-U'}{2} \ .$$

Then

$$X' = A' + \frac{U'-U}{2}$$

and since $X = X'$, we finally obtain

$$X = \tfrac{1}{2}(A + A').$$

(ii) This time

$$G(X) = X'X - I = 0.$$

Hence

$$f^* = \operatorname{tr}[A'-X'][A-X] + \operatorname{tr} U'[XX'-I]$$

so that

$$\frac{\partial f^*}{\partial X} = -2A + 2X + [U+U']\,X$$
$$= 0 \quad \text{for} \quad X = A - \left[\frac{U+'U'}{2}\right] X.$$

Premultiplying by X' and using the condition

$$X'X = I$$

we obtain

$$X'A = I + \frac{U+U'}{2}$$

and on transposing

$$A'X = I + \frac{U+U'}{2} \ .$$

Hence

$$A'X = X'A. \tag{7.14}$$

If a solution to (7.14) exists, there are various ways of solving this matrix equation.

For example with the help of (2.13) and Example (2.7) we can write it as

$$[(I \otimes A') - (A' \otimes I)U]\ \mathbf{x} = \mathbf{0} \tag{7.15}$$

where U is a permutation matrix (see (2.24)) and

$$\mathbf{x} = \text{vec}\, X\,.$$

We have now reduced the matrix equation into a system of homogeneous equations which can be solved by a standard method.

If a non-trivial solution to (7.15) does exist, it is not unique. We must scale it appropriately for X to be orthogonal.

There may, of course, be more than one linearly independent solution to (7.15). We must choose the solution corresponding to X being an orthogonal matrix.

Example 7.2

Given

$$A = \begin{bmatrix} 1 & 2 \\ -1 & 1 \end{bmatrix},$$

find the othogonal matrix X which is the least squares best approximation to A.

Solution

$$[I \otimes A'] = \begin{bmatrix} 1 & -1 & 0 & 0 \\ 2 & 1 & 0 & 0 \\ 0 & 0 & 1 & -1 \\ 0 & 0 & 2 & 1 \end{bmatrix} \quad \text{and} \quad [A' \otimes I]\,U = \begin{bmatrix} 1 & -1 & 0 & 0 \\ 0 & 0 & 1 & -1 \\ 2 & 1 & 0 & 0 \\ 0 & 0 & 2 & 1 \end{bmatrix}.$$

Equation (7.15) can now be written as

$$\begin{bmatrix} 0 & 0 & 0 & 0 \\ 2 & 1 & -1 & 1 \\ -2 & -1 & 1 & -1 \\ 0 & 0 & 0 & 0 \end{bmatrix} \mathbf{x} = \mathbf{0}\,.$$

There are 3 non-trivial (linearly independent) solutions, (see [18] p. 131). They are

$$\mathbf{x} = [1\ -2\ 1\ 1]', \quad \mathbf{x} = [1\ 1\ 2\ -1]' \quad \text{and} \quad \mathbf{x} = [2\ -3\ 3\ 2]'.$$

Only the last solution leads to an orthogonal matrix X, it is

$$X = \frac{1}{\sqrt{13}} \begin{bmatrix} 2 & 3 \\ -3 & 2 \end{bmatrix}.$$

7.4 PROBLEM 2 – THE GENERAL LEAST SQUARES PROBLEM

The linear regression problem presents itself in the following form:

N samples from a population are considered. The ith sample consists of an observation from a variable Y and observations from variables $X_1, X_2, \ldots, X_n$ (say).

We assume a linear relationship between the variables. If the variables are measured from zero, the relationship is of the form

$$y_i = b_0 + b_1 x_{i1} + b_2 x_{i2} + \ldots + b_n x_{in} + e_i\,. \tag{7.16}$$

If the observations are measured from their means over the N samples, then

$$y_i = b_1 x_{i1} + b_2 x_{i2} + \ldots + b_n x_{in} + e_i \quad (i = 1, 2, \ldots N) \tag{7.17}$$

$b_0, b_1, b_2, \ldots, b_n$ are estimated parameters and e_i is the corresponding residual.

In matrix notation we can write the above equations as

$$\mathbf{y} = X\mathbf{b} + \mathbf{e} \tag{7.18}$$

where

$$\mathbf{y} = \begin{bmatrix} y_1 \\ y_2 \\ \vdots \\ y_N \end{bmatrix}, \quad \mathbf{b} = \begin{bmatrix} b_1 \\ b_2 \\ \vdots \\ b_n \end{bmatrix}, \quad \mathbf{e} = \begin{bmatrix} e_1 \\ e_2 \\ \vdots \\ e_N \end{bmatrix}$$

and

$$X = \begin{bmatrix} 1 & x_{12} & \ldots & x_{1n} \\ 1 & x_{22} & \ldots & x_{2n} \\ \vdots & \vdots & & \vdots \\ 1 & x_{N2} & \ldots & x_{Nn} \end{bmatrix} \quad \text{or} \quad X = \begin{bmatrix} x_{11} & x_{12} & \ldots & x_{1n} \\ x_{21} & x_{22} & \ldots & x_{2n} \\ \vdots & & & \\ x_{N1} & x_{N2} & \ldots & x_{Nn} \end{bmatrix}.$$

As already indicated, the 'goodness of fit' criterion is the minimisation with respect to the parameters $\mathbf{b}$ of the sum of the squares of the residuals, which in this case is

$$S = \mathbf{e}'\mathbf{e} = (\mathbf{y}' - \mathbf{b}'X')(\mathbf{y} - X\mathbf{b})\,.$$

Making use of the results in table (4.4), we obtain

$$\begin{aligned} \frac{\partial(\mathbf{e}'\mathbf{e})}{\partial\mathbf{b}} &= -(\mathbf{y}'X)' - X'\mathbf{y} + (X'X\mathbf{b} + X'X\mathbf{b}) \\ &= -2X'\mathbf{y} + 2X'X\mathbf{b} \\ &= 0 \quad \text{for} \quad X'X\hat{\mathbf{b}} = X'\mathbf{y}\,. \end{aligned} \tag{7.19}$$

where $\hat{\mathbf{b}}$ is the least squares estimate of $\mathbf{b}$.

If $(X'X)$ is non-singular, we obtain from (7.19)

$$\hat{\mathbf{b}} = (X'X)^{-1}X'\mathbf{y}\,. \tag{7.20}$$

We can write (7.19) as

$$X'(\mathbf{y} - X\hat{\mathbf{b}}) = \mathbf{0}$$

or

$$X'\mathbf{e} = \mathbf{0} \tag{7.21}$$

which is the matrix form of the normal equations defined in section 7.2.

Example 7.3

Obtain the normal equations for a least squares approximation when each sample consists of one observation from Y and one observation from

(i) a random variable X
(ii) two random variables X and Z.

Solution

(i)

$$X = \begin{bmatrix} 1 & x_1 \\ 1 & x_2 \\ \vdots & \vdots \\ 1 & x_N \end{bmatrix}, \quad \mathbf{y} = \begin{bmatrix} y_1 \\ y_2 \\ \vdots \\ y_N \end{bmatrix}, \quad \hat{\mathbf{b}} = \begin{bmatrix} \hat{b}_1 \\ \hat{b}_2 \end{bmatrix},$$

hence

$$X'[\mathbf{y} - X\hat{\mathbf{b}}] = \begin{bmatrix} \Sigma y_i - \hat{b}_1 N - \hat{b}_2 \,\Sigma x_i \\ \Sigma x_i y_i - \hat{b}_1 \,\Sigma x_i - \hat{b}_2 \,\Sigma x_i^2 \end{bmatrix}.$$

So that the normal equations are

$$\Sigma y_i = \hat{b}_1 N + \hat{b}_2 \,\Sigma x_i$$

and

$$\Sigma x_i y_i = \hat{b}_1 \,\Sigma x_i + \hat{b}_2 \,\Sigma x_i^2 .$$

(ii) In this case

$$X = \begin{bmatrix} 1 & x_1 & z_1 \\ 1 & x_2 & z_2 \\ \vdots & \vdots & \vdots \\ 1 & x_N & z_N \end{bmatrix}, \quad \mathbf{y} = \begin{bmatrix} y_1 \\ y_2 \\ \vdots \\ y_N \end{bmatrix}, \quad \hat{\mathbf{b}} = \begin{bmatrix} \hat{b}_1 \\ \hat{b}_2 \\ \hat{b}_3 \end{bmatrix}.$$

The normal equations are

$$\Sigma y_i = \hat{b}_1 N + \hat{b}_2 \,\Sigma x_i + \hat{b}_3 \,\Sigma z_i$$

$$\Sigma x_i y_i = \hat{b}_1 \,\Sigma x_i + \hat{b}_2 \,\Sigma x_i^2 + \hat{b}_3 \,\Sigma x_i z_i$$

and

$$\Sigma x_i z_i = \hat{b}_1 \,\Sigma z_i + \hat{b}_2 \,\Sigma x_i z_i + \hat{b}_3 \,\Sigma z_i^2 .$$

7.5 PROBLEM 3 – MAXIMUM LIKELIHOOD ESTIMATE OF THE MULTIVARIATE NORMAL

Let $X_i (i = 1, 2, \ldots, n)$ be n random variables each having a normal distribution with mean μ_i and standard deviation σ_i, that is

$$X_i = n(\mu_i, \sigma_i)\,. \tag{7.22}$$

The **joint probability density function** (p.d.f.) of the n random variables is

$$f(x_1, x_2, \ldots, x_n) = \frac{1}{(2\pi)^{n/2}|V|^{1/2}} \exp\left(\frac{-\frac{1}{2}(\mathbf{x}-\boldsymbol{\mu})'V^{-1}(\mathbf{x}-\boldsymbol{\mu})}{2}\right) \tag{7.23}$$

where

$$-\infty < x_i < \infty \qquad (i = 1, 2, \ldots, n)$$

and

$$V = \begin{bmatrix} \sigma_{11} & \sigma_{12} & \ldots & \sigma_{1n} \\ \sigma_{12} & \sigma_{22} & \ldots & \sigma_{2n} \\ \vdots & \vdots & & \vdots \\ \sigma_{1n} & \sigma_{2n} & \ldots & \sigma_{nn} \end{bmatrix}$$

is the **covariance** matrix.

$$\boldsymbol{\mu}' = [\mu_1, \mu_2, \ldots, \mu_n], \qquad \mathbf{x}' = [x_1, x_2, \ldots, x_n]$$

and

$$\sigma_{ij} = \rho_{ij}\sigma_i\sigma_j \quad (i \neq j)$$
$$\sigma_{ii} = \sigma_i^2$$

are the covariances of the random variables.

ρ_{ij} is the correlation coefficient between X_i and X_j. The covariance matrix V is symmetric and positive definite.

Equation (7.23) is called a **multivariate normal** p.d.f. Maximum likelihood estimates have certain properties (for example, they are asymptotically efficient) which makes them very useful in estimation and hypothesis testing problems.

For a sample of N observations from the multivariate normal distribution (7.23) the likelihood function is

$$L = \frac{1}{(2\pi)^{nN/2}|V|^{N/2}} \exp\left\{-\frac{1}{2}\sum_{i=1}^{N}(\mathbf{x}_i-\boldsymbol{\mu})'V^{-1}(\mathbf{x}_i-\boldsymbol{\mu})\right\}$$

so that

$$\log L = C - \frac{N}{2}\log|V| - \frac{1}{2}\sum_{i=1}^{N}(\mathbf{x}_i-\boldsymbol{\mu})'V^{-1}(\mathbf{x}_i-\boldsymbol{\mu}) \tag{7.24}$$

where C is a constant.

(a) The maximum likelihood estimate of $\boldsymbol{\mu}$

On expanding the last term of (7.24), we obtain

$$-\frac{1}{2}\sum_{i=1}^{N}\{\mathbf{x}_i{}'V^{-1}\mathbf{x}_i-\boldsymbol{\mu}'V^{-1}\mathbf{x}_i-\mathbf{x}_i{}'V^{-1}\boldsymbol{\mu}+\boldsymbol{\mu}'V^{-1}\boldsymbol{\mu}\}.$$

With the help of table (4.4) and using the result

$$(\mathbf{x}_i{}'V^{-1})' = V^{-1}\mathbf{x}_i \qquad \text{(since } V \text{ is symmetric)}$$

we differentiate with respect to $\boldsymbol{\mu}$, to obtain

$$\frac{\partial \log L}{\partial \boldsymbol{\mu}} = V^{-1}\sum_{i=1}^{N}(\mathbf{x}_i-\boldsymbol{\mu})$$

$$= 0 \quad \text{when} \quad \hat{\boldsymbol{\mu}} = \frac{\Sigma \mathbf{x}_i}{N} = \bar{\mathbf{x}}\,.$$

Hence the maximum likelihood estimate of $\boldsymbol{\mu}$ is $\boldsymbol{\mu} = \bar{\mathbf{x}}$, the sample mean.

(b) The maximum likelihood estimate of V

We note the following results:

(1) $$\sum_{i=1}^{N}\mathbf{y}_i'V^{-1}\mathbf{y}_i = \operatorname{tr}(Y'V^{-1}Y) = \operatorname{tr}(YY'V^{-1})$$

where $Y = [\mathbf{y}_1\ \mathbf{y}_2 \ldots \mathbf{y}_N]$

and $\mathbf{y}_i = \mathbf{x}_i - \boldsymbol{\mu} \quad (i = 1, 2, \ldots, N.$

V^{-1} is a symmetric matrix.

(2) By Example 5.3, but taking account of the symmetry of V^{-1} (see Example 4.4)

$$\frac{\partial \log |V^{-1}|}{\partial V^{-1}} = 2V - \operatorname{diag}\{V\}\,.$$

(3) If X is a symmetric matrix

$$\frac{\partial \operatorname{tr}(AX)}{\partial X} = A + A' - \operatorname{diag}\{A\}\,.$$

Let $A = YY'$ and $X = V^{-1}$, then

$$\frac{\partial \operatorname{tr}(YY'V^{-1})}{\partial V^{-1}} = 2YY' - \operatorname{diag}\{YY'\}\,.$$

We now write (7.24) as

$$\log L = C + \frac{N}{2}\log V^{-1} - \frac{1}{2}\operatorname{tr}(YY'V^{-1}) .$$

Differentiating log L with respect to V^{-1}, using the estimate $\boldsymbol{\mu} = \bar{\mathbf{x}}$, and the results (2) and (3) above, we obtain

$$\frac{\partial \log L}{\partial V^{-1}} = \frac{N}{2}[2V - \operatorname{diag}\{V\}] - YY' + \frac{1}{2}\operatorname{diag}\{YY'\} .$$

Let $Q = NV - YY'$, then

$$\frac{\partial \log L}{\partial V^{-1}} = Q - \frac{1}{2}\operatorname{diag}\{Q\}$$

$$= 0 \quad \text{when} \quad 2Q = \operatorname{diag}\{Q\} .$$

Since Q is symmetric, the only solution to the above equation is

$$Q = 0.$$

It follows that the maximum likelihood estimate of V is

$$\hat{V} = \frac{\Sigma(\mathbf{x}_i - \bar{\mathbf{x}})(\mathbf{x}_i - \bar{\mathbf{x}})'}{N} .$$

7.6 PROBLEM 4 – EVALUATION OF THE JACOBIANS OF SOME TRANSFORMATIONS

The interest in Jacobians arises from their importance particularly with reference to a change of variables in multiple integration.

In terms of scalars, the problem presents itself in the following way.

We consider a multiple integral of a subset R of an n-dimensional space

$$\int_R f(x_1, x_2, \ldots, x_n)\, dx_1\, dx_2 \ldots, dx_n. \tag{7.25}$$

where f is a piecewise continuous function in R.

We consider a one to one transformation which maps R onto a subset T

$$y_1 = \mu_1(\mathbf{x}), \quad y_2 = \mu_2(\mathbf{x}), \quad \ldots, \quad y_n = \mu_n(\mathbf{x}) \tag{7.26}$$

and the inverse transformation

$$x_1 = w_1(\mathbf{y}),\ x_2 = w_2(\mathbf{y}),\ \ldots,\ x_n = w_n(\mathbf{y}) \tag{7.27}$$

where

$$\mathbf{x}' = [x_1, x_2, \ldots, x_n] \quad \text{and} \quad \mathbf{y}' = [y_1, y_2, \ldots, y_n] .$$

Assuming the first partial derivations of the inverse transformation (7.27) to be continuous, (7.25) can be expressed as

$$\int_T f[w_1(\mathbf{y}), w_2(\mathbf{y}), \ldots, w_n(\mathbf{y})]\, |J|\, dy_1\, dy_2 \ldots dy_n \tag{7.28}$$

where $|J|$ can be written as

$$|J| = \begin{vmatrix} \frac{\partial x_1}{\partial y_1} & \frac{\partial x_2}{\partial y_1} & \cdots & \frac{\partial x_n}{\partial y_1} \\ \frac{\partial x_1}{\partial y_2} & \frac{\partial x_2}{\partial y_2} & \cdots & \frac{\partial x_n}{\partial y_2} \\ \vdots & \vdots & & \vdots \\ \frac{\partial x_1}{\partial y_n} & \frac{\partial x_2}{\partial y_n} & \cdots & \frac{\partial x_n}{\partial y_n} \end{vmatrix} = \left| \frac{\partial \mathbf{x}}{\partial \mathbf{y}} \right| , \tag{7.29}$$

subject to $|J|$ not vanishing identically in T.

Example 7.4

Let

$$I = 2\int_R \exp\{-2x_1 + 3x_2\}\, dx_1\, dx_2$$

$$0 < x_1 < \infty, \quad 0 < x_2 < \infty .$$

Consider the transformation

$$y_1 = 2x_1 - x_2$$

$$y_2 = x_2 .$$

Write down the integral corresponding to (7.28).

Solution

We are given

$$R = \{(x_1, x_2):\ 0 < x_1 < \infty,\ 0 < x_2 < \infty\} .$$

The above transformation (corresponding to (7.26)) results in the following inverse transformation (7.27)

$$x_1 = \tfrac{1}{2}(y_1 + y_2)$$

$$x_2 = y_2$$

which defines

$$T = \{(y_1, y_2):\ y_2 > 0,\ y_2 > -y_1,\ -\infty < y_1 < \infty\},$$

and by (7.29)

$$|J| = \begin{vmatrix} \frac{1}{2} & 0 \\ \frac{1}{2} & 1 \end{vmatrix} = \frac{1}{2} .$$

Hence

$$I = \int_T f[\tfrac{1}{2}(y_1 + y_2), y_2]\, dy_1\, dy_2$$

$$= \int_T \exp(-y_1 + 2y_2) dy_1\, dy_2 .$$

Our main interest in this section is to evaluate Jacobians when the transformation corresponding to (7.26) is expressed in matrix form, for example as

$$Y = AXB \tag{7.30}$$

where A, X and B are all assumed to be of order $(n \times n)$.

As in section 5.2 (see (5.1) and (5.2)) we can write (7.6) as

$$\mathbf{y} = P\mathbf{x} \tag{7.31}$$

where $\mathbf{y} = \text{vec}\, Y$, $\mathbf{x} = \text{vec}\, X$ and $P = B' \otimes A$.

In this case

$$\frac{\partial \mathbf{y}}{\partial \mathbf{x}} = B \otimes A'$$

and

$$\frac{\partial \mathbf{x}}{\partial \mathbf{y}} = [B \otimes A']^{-1} = B^{-1} \otimes (A')^{-1} \quad \text{by (2.12)} .$$

It follows that

$$|J| = \left| \frac{\partial \text{ vec } Y}{\partial \text{ vec } X} \right|^{-1} = |B|^{-n}\, |A|^{-n} \text{ (by Property X, p. 27)} \tag{7.32}$$

Example 7.5

Consider the transformation

$$Y = AXB$$

where

$$A = \begin{bmatrix} 2 & -4 \\ -1 & 3 \end{bmatrix} \text{ and } B = \begin{bmatrix} 2 & 1 \\ 1 & 1 \end{bmatrix} .$$

Find the Jacobian of this transformation

(i) By a direct method
(ii) Using (7.32).

Solution

(i) We have

$$X = A^{-1}YB^{-1} = \tfrac{1}{2}\begin{bmatrix} 3y_1+4y_2-3y_3-4y_4 & -3y_1-4y_2+6y_3+8y_4 \\ y_1+2y_2-y_3-2y_4 & -y_1-2y_2+2y_3+4y_4 \end{bmatrix}$$

so that

$$\frac{\partial \mathbf{x}}{\partial \mathbf{y}} = \left(\frac{1}{2}\right)^4 \begin{vmatrix} 3 & 1 & -3 & -1 \\ 4 & 2 & 4 & -2 \\ -3 & -1 & 6 & 2 \\ -4 & -2 & 8 & 4 \end{vmatrix} = \frac{1}{4}\,.$$

(ii) $|A| = 2,\ |B| = 1$ hence $|J| = \frac{1}{4}$.

Similarly, we can use the theory developed in this book to evaluate the Jacobians of many other transformations.

Example 7.6

Evaluate the Jacobian associated with the following transformation

(i) $Y = X^{-1}$
(ii) $Y = X^2$.

Solution

(i) From Example 5.2

$$\frac{\partial \mathbf{y}}{\partial \mathbf{x}} = -X^{-1} \otimes (X^{-1})'$$

so that

$$\frac{\partial \mathbf{x}}{\partial \mathbf{y}} = -X \otimes X'\,.$$

Hence

$$J = \text{mod}\left|\frac{\partial \mathbf{y}}{\partial \mathbf{x}}\right| = |X \otimes X'| = |X|^{-n}\,|X|^{-n} = |X|^{-2n}\,.$$

(ii) From section 4.6

$$\frac{\partial Y}{\partial x_{rs}} = E_{rs}X + XE_{rs}\,,$$

so that by the 2nd transformation principle (see section 5.3)

$$\frac{\partial \mathbf{y}}{\partial \mathbf{x}} = X \otimes I + I \otimes X'$$

and

$$J = |X \otimes I + I \otimes X'|^{-1}$$

7.7 PROBLEM 5 – TO FIND THE DERIVATIVE OF AN EXPONENTIAL MATRIX WITH RESPECT TO A MATRIX

Since we make use of the spectral decomposition of an exponential matrix, we now discuss this technique briefly.

Assume that the matrix $Q = [q_{ij}]$ of order $(n \times n)$ has eigenvalues

$$\lambda_1, \lambda_2, \ldots, \lambda_n$$

(not necessarily distinct) and corresponding linearly independent eigenvectors

$$\mathbf{x}_1, \mathbf{x}_2, \ldots, \mathbf{x}_n .$$

The eigenvectors of Q' are

$$\mathbf{y}_1, \mathbf{y}_2, \ldots, \mathbf{y}_n .$$

These two sets of eigenvectors have the property

$$\mathbf{x}_i' \mathbf{y}_j = 0 \quad \text{or (equivalently)} \quad \mathbf{y}_i' \mathbf{x}_j = 0 \quad (i \neq j) \tag{7.33}$$

and can be normalised so that

$$\mathbf{x}_i' \mathbf{y}_i = 1 \quad \text{or} \quad \mathbf{y}_i' \mathbf{x}_i = 1 \quad (i = 1, 2, \ldots, n) . \tag{7.34}$$

Sets of eigenvectors $\{\mathbf{x}_i\}$ and $\{\mathbf{y}_i\}$ having the properties (7.33) and (7.34) are said to be **properly normalised.**

It is well known (see [18] p. 227) that

$$\exp(Qt) = P \operatorname{diag}\{e^{\lambda_1 t}, e^{\lambda_2 t}, \ldots, e^{\lambda_n t}\} P^{-1}$$

where P is the modal matrix of Q, that is the matrix

$$P = [\mathbf{x}_1 \vdots \mathbf{x}_2 \vdots \ldots \vdots \mathbf{x}_n] . \tag{7.35}$$

It follows from (7.33), (7.34) and (7.35) that

$$P^{-1} = [\mathbf{y}_1, \mathbf{y}_2, \ldots, \mathbf{y}_n]' = \begin{bmatrix} \mathbf{y}_1' \\ \mathbf{y}_2' \\ \vdots \\ \mathbf{y}_n' \end{bmatrix} . \tag{7.36}$$

Hence

$$\exp(Qt) = [\mathbf{x}_1\ \mathbf{x}_2 \ldots \mathbf{x}_n] \begin{bmatrix} e^{\lambda_1 t} & 0 & \ldots & 0 \\ 0 & e^{\lambda_2 t} & \ldots & 0 \\ \vdots & & & \\ 0 & 0 & \ldots & e^{\lambda_n t} \end{bmatrix} \begin{bmatrix} \mathbf{y}_1' \\ \mathbf{y}_2' \\ \vdots \\ \mathbf{y}_n' \end{bmatrix}$$

$$= \mathbf{x}_1\mathbf{y}_1' \exp(\lambda_1 t) + \mathbf{x}_2\mathbf{y}_2' \exp(\lambda_2 t) + \ldots + \mathbf{x}_n\mathbf{y}_n' \exp(\lambda_n t) ,$$

that is

$$\exp(Qt) = \sum_{i=1}^{n} \mathbf{x}_i \mathbf{y}_i' \exp(\lambda_i t) \ . \tag{7.37}$$

The right hand side of (7.37) is known as the **spectral representation** (or **spectral decomposition**) of the exponential matrix $\exp(Qt)$.

We consider a very simple illustrative example.

Example 7.7

Find the spectral representation of the matrix $\exp(Qt)$, where

$$Q = \begin{bmatrix} 3 & 4 \\ -2 & -3 \end{bmatrix} .$$

Solution

$$\lambda_1 = 1,\ \lambda_2 = -1;\ \mathbf{x}_1' = [2\ \ -1],\ \mathbf{x}_2' = [1\ \ -1]$$
$$\mathbf{y}_1' = [1\ \ 1],\ \mathbf{y}_2' = [-1\ \ -2] \ .$$

By (7.37)

$$\exp(Qt) = \begin{bmatrix} 2 \\ -1 \end{bmatrix} [1\ \ 1]\ \exp(t) + \begin{bmatrix} 1 \\ -1 \end{bmatrix} [-1\ \ -2]\ \exp(-t)$$
$$= \begin{bmatrix} 2 & 2 \\ -1 & -1 \end{bmatrix} \exp(t) + \begin{bmatrix} -1 & -2 \\ 1 & 2 \end{bmatrix} \exp(-t) \ .$$

Although we have considered matrices having real eigenvalues, and eigenvectors having real elements, the spectral decomposition (7.37) is also valid for complex elements as can be shown by a slight modification of the above exposition.

By the use of (2.17), that is of the result

$$\exp(I \otimes Q) = I \otimes \exp(Q)$$

we generalise the result (7.37) to

$$\exp(I \otimes Q)t = \Sigma(I \otimes \mathbf{x}_i \mathbf{y}_i') \exp(\lambda_i t) \ . \tag{7.38}$$

We now consider the main problem, to obtain an expression for

$$\partial \Phi / \partial Z$$

where

$$\Phi(t) = \exp(Qt) \ , \tag{7.39}$$

so that

$$\Phi(0) = I \ , \tag{7.40}$$

$$\frac{\mathrm{d}}{\mathrm{d}t} \Phi = Q\Phi \ , \tag{7.41}$$

and

$Z = [z_{ij}]$ is a matrix of order $(r \times s)$.

The matrix Q is assumed to be a function of Z, that is $Q(Z)$. Making use of the result (6.5), we can write

$$\frac{\mathrm{d}}{\mathrm{d}t}\frac{\partial\Phi}{\partial Z} = \frac{\partial(Q\Phi)}{\partial Z} = \frac{\partial Q}{\partial Z}(I\otimes\Phi) + (I\otimes Q)\frac{\partial\Phi}{\partial Z} \tag{7.42}$$

and from (7.40)

$$\frac{\partial\Phi}{\partial Z}(o) = 0\,. \tag{7.43}$$

We next make use of a generalisation of a well known result (see [19] p. 68); Given

$$\frac{\mathrm{d}}{\mathrm{d}t}X = RX + BU$$

and

$$X(o) = 0\,,$$

then

$$X = \int_0^t \exp\{R(t-\tau)\}BU(\tau)\mathrm{d}\tau\,. \tag{7.44}$$

For

$$X = \frac{\partial\Phi}{\partial Z}\,,\quad R = I\otimes Q\,,\quad B = \frac{\partial Q}{\partial Z}\quad\text{and}\quad U = I\otimes\Phi$$

the solution to (7.42) subject to (7.43) becomes

$$\frac{\partial\Phi}{\partial Z} = \int_0^t \exp\{I\otimes Q(t-\tau)\}\frac{\partial Q}{\partial Z}[I\otimes\Phi(t)]\mathrm{d}\tau$$

$$= \int_0^t \sum_{i,j}(I\otimes \mathbf{x}_i\mathbf{y}_i')\exp(\lambda_i(t-\tau))\frac{\partial Q}{\partial Z}[I\otimes \mathbf{x}_j\mathbf{y}_j']\exp(\lambda_j\tau)\,\mathrm{d}\tau$$

(by 7.37 and 7.38)

$$= \sum_{i,j}(I\otimes \mathbf{x}_i\mathbf{y}_i')\frac{\partial Q}{\partial Z}(I\otimes \mathbf{x}_j\mathbf{y}_j')\exp(\lambda_i t)\int_0^t \exp((\lambda_j-\lambda_i)\tau)\,\mathrm{d}\tau\,.$$

Hence,

$$\frac{\partial\Phi}{\partial Z} = \sum_{i,j} I\otimes \mathbf{x}_i\mathbf{y}_i')\frac{\partial Q}{\partial Z}(I\otimes \mathbf{x}_j\mathbf{y}_j')\exp(\lambda_i t)f_{ij}(t) \tag{7.45}$$

where

$$f_{ij}(t) = t \quad\text{if}\quad \lambda_i = \lambda_j$$

and

$$f_{ij}(t) = (1/(\lambda_j-\lambda_i))\,[\exp(\lambda_j-\lambda_i)t)-1]\quad\text{if}\quad \lambda_i\neq\lambda_j\,.$$

Solution to Problems

CHAPTER 1

(1)

$$AB = \begin{bmatrix} A_1.B._1 & A_1.B._2 & A_1.B._3 \\ A_2.B._1 & A_2.B._2 & A_2.B._3 \\ A_3.B._1 & A_3.B._2 & A_3.B._3 \\ A_4.B._1 & A_4.B._2 & A_4.B._3 \end{bmatrix} .$$

(2) (a) The kth column of AE_{ik} is the ith column of A, all other columns are zero.

(b) The ith row of $E_{ik}A$ is the kth row of A, all other rows are zero.

$$AE_{ik} = A\mathbf{e}_i\mathbf{e}_k' = A._i\mathbf{e}_k'$$

$$E_{ik}A = \mathbf{e}_i\mathbf{e}_k'A = \mathbf{e}_iA_k' \quad .$$

(3)

$$\operatorname{tr} ABC = \sum_i \mathbf{e}_i'ABC\mathbf{e}_i = \sum_i (\mathbf{e}_i'A)B(C\mathbf{e}_i)$$

$$= \sum_i A_i'.BC._i \quad .$$

(4)

$$\operatorname{tr} AE_{ij} = \sum_k \mathbf{e}_k' E_{ij}\mathbf{e}_k = \sum_{k,r,s} \mathbf{e}_k' a_{rs}E_{rs}E_{ij}\mathbf{e}_k$$

$$= \sum_{k,r,s} a_{rs}\mathbf{e}_k'\mathbf{e}_r\mathbf{e}_s'\mathbf{e}_i\mathbf{e}_j'\mathbf{e}_k$$

$$= \sum_{k,r,s} a_{rs}\delta_{kr}\delta_{si}\delta_{jk} = a_{ji}.$$

(5) $$A = \sum_{i,j} a_{ij} A_{ij} = \sum_{i,j} \operatorname{tr}(BE_{ij}\delta_{ij}) E_{ij}$$

$$= \sum_{i,j,k} \mathbf{e}_k' BE_{ii}\mathbf{e}_k E_{ii} = \sum_{i,j,k} \mathbf{e}_k' B\mathbf{e}_i \delta_{ik} E_{ii}$$

$$= \sum_i \mathbf{e}_i' B\mathbf{e}_i E_{ii} = \sum_i b_{ii} E_{ii} = \operatorname{diag}\{B\}.$$

CHAPTER 2

(1) Since U is an orthogonal matrix, the result follows.
More formally,

$$\sum_{r,s} [E_{rs}(m \times n) \otimes E_{sr}(n \times m)]\,[E_{sr}(n \times m) \otimes E_{rs}(m \times n)]$$

$$= \sum_{r,s} [E_{rs}(m \times n) E_{sr}(n \times m)] \otimes [E_{sr}(n \times m) E_{rs}(m \times n)]$$

$$= \sum_{r,s} [\delta_{ss} E_{rr}(m \times m)] \otimes [\delta_{rr} E_{ss}(n \times n)]$$

$$= \sum_r E_{rr}(m \times m)] \otimes [\sum_s E_{ss}(n \times n)]$$

$$= I_m \otimes I_n = I_{mn},$$ the result follows.

(3) (a)

$$A \otimes B = \begin{bmatrix} -2 & 2 & -1 & 1 \\ 4 & 0 & 2 & 0 \\ 0 & 0 & -1 & 1 \\ 0 & 0 & 2 & 0 \end{bmatrix}, \quad B \otimes A = \begin{bmatrix} -2 & -1 & 2 & 1 \\ 0 & -1 & 0 & 1 \\ 4 & 2 & 0 & 0 \\ 0 & 2 & 0 & 0 \end{bmatrix}$$

(b)

$$U_1 = U_2 = \begin{bmatrix} 1 & 0 & 0 & 0 \\ 0 & 0 & 1 & 0 \\ 0 & 1 & 0 & 0 \\ 0 & 0 & 0 & 1 \end{bmatrix}.$$

(4) See [18] p. 228 for methods of calculating matrix exponentials.

(a)

$$\exp(A) = \begin{bmatrix} 2e - e^{-1} & 2(e - e^{-1}) \\ e^{-1} - e & 2e^{-1} - e \end{bmatrix}$$

(b)

$$\exp(A \otimes I) = \begin{bmatrix} 2e - e^{-1} & 0 & 2(e - e^{-1}) & 0 \\ 0 & 2e - e^{-1} & 0 & 2(e - e^{-1}) \\ -(e - e^{-1}) & 0 & -e + 2e^{-1} & 0 \\ 0 & -(e - e^{-1}) & 0 & -e + 2e^{-1} \end{bmatrix}$$

$$\exp(A) \otimes I = \begin{bmatrix} 2e - e^{-1} & 0 & 2(e - e^{-1}) & 0 \\ 0 & 2e - e^{-1} & 0 & 2(e - e^{-1}) \\ e^{-1} - e & 0 & 2e^{-1} - e & 0 \\ 0 & e^{-1} - e & 0 & 2e^{-1} - e \end{bmatrix} .$$

Hence $\exp(A) \otimes I = \exp(A \otimes I)$.

(5) (a)

$$A^{-1} = \begin{bmatrix} 1 & 1 \\ -1 & -2 \end{bmatrix}, \quad B^{-1} = \frac{1}{2}\begin{bmatrix} -4 & 2 \\ 3 & -1 \end{bmatrix}, \quad \text{so that}$$

$$A^{-1} \otimes B^{-1} = \frac{1}{2}\begin{bmatrix} -4 & 2 & -4 & 2 \\ 3 & -1 & 3 & -1 \\ 4 & -2 & 8 & -4 \\ -3 & 1 & -6 & 2 \end{bmatrix} .$$

(b) As

$$A \otimes B = \begin{bmatrix} 2 & 4 & 1 & 2 \\ 6 & 8 & 3 & 4 \\ -1 & -2 & -1 & -2 \\ -3 & -4 & -3 & -4 \end{bmatrix}, \quad \text{it follows that}$$

$$(A \otimes B)^{-1} = \begin{bmatrix} -2 & 1 & -2 & 1 \\ 3/2 & -1/2 & 3/2 & -1/2 \\ 2 & -1 & 4 & -2 \\ -3/2 & 1/2 & -3 & -4 \end{bmatrix}$$

This verifies (2.12)

(6) (a) For A; $\lambda_1 = -1$, $\lambda_2 = 2$, $\mathbf{x}_1' = [1 \ 4]$ and $\mathbf{x}_2' = [1 \ 1]$.

For B; $\mu_1 = 1$, $\mu_2 = 4$, $\mathbf{y}_1' = [1 \ -1]$ and $\mathbf{y}_2' = [1 \ 2]$.

(b)

$$A \otimes B = \begin{bmatrix} 6 & 3 & -2 & -1 \\ 6 & 9 & -2 & -3 \\ 8 & 4 & -4 & -2 \\ 8 & 12 & -4 & -6 \end{bmatrix} = E \text{ (say)} .$$

$$C(\lambda) = |\lambda I - E| = \lambda^4 - 5\lambda^3 - 30\lambda^2 + 40\lambda - 30$$
$$= (\lambda+1)(\lambda+4)(\lambda-2)(\lambda-8)\,.$$

Hence the eigenvalues of E are

$$\{-1, -4, 2, 8\} = \{\lambda_1\mu_1, \lambda_1\mu_2, \lambda_2\mu_1, \lambda_2\mu_2\}.$$

The corresponding eigenvectors of E are:

$$\begin{bmatrix}1\\-1\\4\\-4\end{bmatrix}, \begin{bmatrix}1\\2\\4\\8\end{bmatrix}, \begin{bmatrix}1\\-1\\1\\-1\end{bmatrix} \quad \text{and} \quad \begin{bmatrix}1\\2\\1\\2\end{bmatrix}.$$

(c) This verifies Property IX

(7) For some non-singular P and Q

$$A = P^{-1}CP \quad \text{and} \quad B = Q^{-1}DQ\,.$$

Hence

$$\begin{aligned} A \otimes B &= P^{-1}CP \otimes Q^{-1}DQ \\ &= (P^{-1} \otimes Q^{-1})(CP \otimes DQ) && \text{by (2.11)} \\ &= (P \otimes Q)^{-1}(C \otimes D)(P \otimes Q) && \text{by (2.12) and (2.11)} \\ &= R^{-1}(C \otimes D)R \end{aligned}$$

where

$$R = P \otimes Q\,.$$

The result follows.

CHAPTER 4

(1)

$$\frac{\partial y}{\partial X} = \begin{bmatrix} 2x_{22} & 0 & -x_{21} \\ -x_{13} & 2x_{11} & 0 \end{bmatrix}, \quad \frac{\partial Y}{\partial x} = \begin{bmatrix} 1 & 2e^{2x} \\ 0 & -x^{-2} \\ 4x & \cos x \end{bmatrix}$$

(2) (a)

$$|X| = \exp(x)\sin x - x\cos x,$$

$$\frac{\partial |X|}{\partial X} = \begin{bmatrix} x & -2e^{2x} \\ -2e^{2x} & \sin x \end{bmatrix}.$$

$$|X| = x\sin x - \exp(2x)$$

$$\frac{\partial |X|}{\partial X} = \begin{bmatrix} e^{x} & -\cos x \\ -x & \sin x \end{bmatrix}.$$

(b) $$\frac{\partial |X|}{\partial X} = \begin{bmatrix} X_{11} & X_{12} \\ X_{21} & X_{22} \end{bmatrix} = \begin{bmatrix} e^x & -\cos x \\ -x & \sin x \end{bmatrix}.$$

$$\frac{\partial |X|}{\partial X} = 2[X_{ij}] - \text{diag}\,\{X_{ij}\}$$

$$= \begin{bmatrix} 2x & -2e^x \\ -2e^x & 2\sin x \end{bmatrix} - \begin{bmatrix} x & 0 \\ 0 & \sin x \end{bmatrix}.$$

(3) $$Y = \begin{bmatrix} x_{11}^2 + x_{21}^2 & x_{11}x_{12} + x_{21}x_{22} & x_{11}x_{13} + x_{21}x_{23} \\ x_{12}x_{11} + x_{22}x_{21} & x_{12}^2 + x_{22}^2 & x_{12}x_{13} + x_{22}x_{23} \\ x_{13}x_{11} + x_{23}x_{21} & x_{13}x_{12} + x_{23}x_{22} & x_{13}^2 + x_{23}^2 \end{bmatrix},$$

hence

(a) $$\frac{\partial Y}{\partial x_{21}} = \begin{bmatrix} 2x_{21} & x_{22} & x_{23} \\ x_{22} & 0 & 0 \\ x_{23} & 0 & 0 \end{bmatrix},$$

From Example 4.8

$$\frac{\partial Y}{\partial x_{21}} = E'_{21}X + X'E_{21} =$$

$$\begin{bmatrix} 0 & 1 \\ 0 & 0 \\ 0 & 0 \end{bmatrix} \begin{bmatrix} x_{11} & x_{12} & x_{13} \\ x_{21} & x_{22} & x_{23} \end{bmatrix} + \begin{bmatrix} x_{11} & x_{21} \\ x_{12} & x_{22} \\ x_{13} & x_{23} \end{bmatrix} \begin{bmatrix} 0 & 0 & 0 \\ 1 & 0 & 0 \end{bmatrix}$$

(b) Since $y_{13} = x_{11}x_{13} + x_{21}x_{23}$

$$\frac{\partial y_{13}}{\partial X} = \begin{bmatrix} x_{13} & 0 & x_{11} \\ x_{23} & 0 & x_{21} \end{bmatrix} =$$

$$\begin{bmatrix} x_{11} & x_{12} & x_{13} \\ x_{21} & x_{22} & x_{23} \end{bmatrix} \begin{bmatrix} 0 & 0 & 0 \\ 0 & 0 & 0 \\ 1 & 0 & 0 \end{bmatrix} + \begin{bmatrix} x_{11} & x_{12} & x_{13} \\ x_{21} & x_{22} & x_{23} \end{bmatrix} \begin{bmatrix} 0 & 0 & 1 \\ 0 & 0 & 0 \\ 0 & 0 & 0 \end{bmatrix}$$

which is the result in Example 4.8.

(4) (a)

$$\frac{\partial Y}{\partial x_{rs}} = E_{rs}AX + XAE_{rs} ,$$

$$\frac{\partial y_{ij}}{\partial X} = E_{ij}X'A' + A'X'E_{ij} .$$

(b)

$$\frac{\partial Y}{\partial x_{rs}} = E'_{rs}AX' + X'AE'_{rs} ,$$

$$\frac{\partial y_{ij}}{\partial X} = AX'E'_{ij} + E'_{ij}X'A .$$

(5) By (4.10)

$$\begin{aligned}\frac{\partial |Y|}{\partial x_{rs}} &= \text{tr}\,\{|Y|(Y^{-1})'B'E'_{rs}A'\} \\ &= |Y|\,\text{tr}\,\{A'(Y^{-1})'B'E'_{rs}\} \\ &= |Y|\,(\text{vec}\,E_{rs})'\,\text{vec}\,[A'(Y^{-1})'B'] \\ &= |AXB|\,z_{rs}\end{aligned}$$

where

$$[z_{rs}] = Z = A'[(AXB)^{-1}]'B' .$$

(6) (a) Since

$$\frac{\partial (XX')}{\partial x_{rs}} = E_{rs}X' + XE'_{rs} ,$$

$$\frac{\partial Y}{\partial x_{rs}} = E_{rs}(X')^2 + XE'_{rs}X' + XX'E'_{rs}$$

(b)

$$\frac{\partial Y}{\partial x_{rs}} = E'_{rs}X'X + X'E'_{rs}X + (X')^2E_{rs} .$$

CHAPTER 5

(1) Since

$$\begin{bmatrix} y_{11} \\ y_{21} \\ y_{12} \\ y_{22} \end{bmatrix} = \begin{bmatrix} a_{11}x_{11} + a_{12}x_{12} \\ a_{21}x_{11} + a_{22}x_{12} \\ a_{11}x_{21} + a_{12}x_{22} \\ a_{21}x_{21} + a_{22}x_{22} \end{bmatrix} ,$$

$$\frac{\partial \operatorname{vec} Y}{\partial \operatorname{vec} X} = \begin{bmatrix} a_{11} & a_{21} & 0 & 0 \\ 0 & 0 & a_{11} & a_{21} \\ a_{12} & a_{22} & 0 & 0 \\ 0 & 0 & a_{12} & a_{22} \end{bmatrix} .$$

(2) (a) $$\frac{\partial \operatorname{vec} Y}{\partial \operatorname{vec} X} = (B \otimes A')_{(n)} \quad \text{by (5.18)}$$

(b) $$\frac{\partial \operatorname{vec} Y}{\partial \operatorname{vec} X} = X \otimes I + I \otimes X' .$$

(3) (a) $$\frac{\partial \operatorname{tr} Y}{\partial x_{rs}} = \operatorname{tr} AE_{rs}B = \operatorname{tr} E'_{rs}A'B' = (\operatorname{vec} E_{rs})' (\operatorname{vec} A'B') ,$$

hence

$$\frac{\partial \operatorname{tr} Y}{\partial X} = A'B' .$$

(b) $$\frac{\partial \operatorname{tr} Y}{\partial x_{rs}} = 2 \operatorname{tr} E'_{rs} X' ,$$

hence

$$\frac{\partial \operatorname{tr} Y}{\partial X} = 2X' .$$

(c) $$\frac{\partial \operatorname{tr} Y}{\partial x_{rs}} = 2 \operatorname{tr} E'_{rs} X ,$$

hence

$$\frac{\partial \operatorname{tr} Y}{\partial X} = 2X .$$

(4) (a) $$\frac{\partial \operatorname{tr} Y}{\partial x_{rs}} = -\operatorname{tr} X^{-1} E_{rs} X^{-1} = -\operatorname{tr} E'_{rs} (X^{-2})' ,$$

hence

$$\frac{\partial \operatorname{tr} Y}{\partial X} = -(X^{-2})' .$$

(b) $$\frac{\partial \operatorname{tr} Y}{\partial x_{rs}} = -\operatorname{tr} AX^{-1} E_{rs} X^{-1} B ,$$

hence

$$\frac{\partial \operatorname{tr} Y}{\partial X} = -(X^{-1}BAX^{-1})' .$$

(c)
$$\frac{\partial \operatorname{tr} Y}{\partial x_{rs}} = \operatorname{tr} E_{rs} X^{n-1} + \operatorname{tr} X E_{rs} X^{n-2} + \ldots + \operatorname{tr} X^{n-1} E_{rs}$$
hence
$$\frac{\partial \operatorname{tr} Y}{\partial x_{rs}} = n(X^{n-1})'$$

(d)
$$\exp(X) = I + X + \frac{1}{2!}X^2 + \frac{1}{3!}X^3 + \ldots$$
hence by the result (c) above
$$\frac{\partial \operatorname{tr} Y}{\partial X} = \exp(X').$$

(5) (a) (i)
$$dY = \begin{bmatrix} a_{11}dx_{11} + a_{12}dx_{21} & a_{11}dx_{12} + a_{12}dx_{22} \\ a_{21}dx_{11} + a_{22}dx_{21} & a_{21}dx_{12} + a_{22}dx_{22} \end{bmatrix}$$
$$= \begin{bmatrix} a_{11} & a_{12} \\ a_{21} & a_{22} \end{bmatrix} \begin{bmatrix} dx_{11} & dx_{12} \\ dx_{21} & dx_{22} \end{bmatrix} = A(dX).$$

(ii)
$$dY = \begin{bmatrix} 2x_{11}dx_{11} + 2x_{21}dx_{21} & x_{11}dx_{12} + x_{12}dx_{11} + x_{21}dx_{22} + x_{22}dx_{21} \\ x_{11}dx_{12} + x_{12}dx_{11} + x_{22}dx_{21} + d_{21}dx_{22} & 2x_{12}dx_{12} + 2x_{22}dx_{22} \end{bmatrix}$$
$$= \begin{bmatrix} dx_{11} & dx_{21} \\ dx_{12} & dx_{22} \end{bmatrix} \begin{bmatrix} x_{11} & x_{12} \\ x_{21} & x_{22} \end{bmatrix} + \begin{bmatrix} x_{11} & x_{21} \\ x_{12} & x_{22} \end{bmatrix} \begin{bmatrix} dx_{11} & dx_{12} \\ dx_{21} & dx_{22} \end{bmatrix}$$
$$= (dX)'X + X'(dX).$$

(iii)
$$dY = \begin{bmatrix} 2x_{11}dx_{11} + x_{12}dx_{21} + x_{21}dx_{12} & x_{11}dx_{12} + x_{12}dx_{11} + x_{12}dx_{22} + x_{22}dx_{12} \\ x_{11}dx_{21} + x_{21}dx_{11} + x_{22}dx_{21} + x_{21}dx_{22} & x_{21}dx_{12} + x_{12}dx_{21} + 2x_{22}dx_{22} \end{bmatrix}$$
$$= \begin{bmatrix} x_{11}dx_{11} + x_{12}dx_{21} & x_{11}dx_{12} + x_{12}dx_{22} \\ x_{21}dx_{11} + x_{22}dx_{21} & x_{21}dx_{12} + x_{22}dx_{22} \end{bmatrix} \begin{bmatrix} x_{11}dx_{11} + x_{21}dx_{12} & x_{12}dx_{11} + x_{22}dx_{12} \\ x_{11}dx_{21} + x_{21}dx_{21} & x_{12}dx_{21} + x_{22}dx_{22} \end{bmatrix}$$
$$= X(dX) + (dX)X.$$

(b) Write $Y = UV$ where $U = AX$ and $V = BX$,
then $dY = U(dV) + (dU)V$
$\qquad = AXB(dX) + A(dX)BX$.

CHAPTER 6

(1)
$$\frac{\partial Y}{\partial X} = \begin{bmatrix} 0 & x_{21} & -\sin(x_{12}+x_{22}) & 0 \\ x_{11}e^{x_{11}x_{12}} & 0 & x_{12}e^{x_{11}x_{12}} & x_{22} \\ -\sin(x_{12}+x_{22}) & x_{11} & 0 & 0 \\ 0 & 0 & 0 & x_{12} \end{bmatrix}$$

(2)
$$\frac{\partial X}{\partial X_{11}} = \begin{bmatrix} 1 & 0 \\ 0 & 0 \\ 0 & 0 \end{bmatrix}, \quad \frac{\partial X}{\partial x_{12}} = \begin{bmatrix} 0 & 0 \\ 1 & 0 \\ 0 & 0 \end{bmatrix}, \quad \frac{\partial X}{\partial x_{13}} = \begin{bmatrix} 0 & 0 \\ 0 & 0 \\ 1 & 0 \end{bmatrix} \quad \text{and so on ,}$$

hence by (6.1)

$$\frac{\partial X}{\partial X} = \begin{bmatrix} 1 & 0 & 0 & 1 \\ 0 & 0 & 0 & 0 \\ 0 & 0 & 0 & 0 \\ 0 & 0 & 0 & 0 \\ 1 & 0 & 0 & 1 \\ 0 & 0 & 0 & 0 \\ 0 & 0 & 0 & 0 \\ 0 & 0 & 0 & 0 \\ 1 & 0 & 0 & 1 \end{bmatrix} = \bar{U}.$$

(3) Since
$$X^{-1} = \frac{1}{\Delta}\begin{bmatrix} x_{22} & -x_{12} \\ -x_{21} & x_{11} \end{bmatrix}$$

where $\Delta = x_{11}x_{22} - x_{12}x_{21}$, we can calculate $\partial X^{-1}/\partial x_{rs}$, for example

$$\frac{\partial X^{-1}}{\partial x_{11}} = \frac{1}{\Delta^2}\begin{bmatrix} -x_{22}^2 & x_{12}x_{22} \\ x_{21}x_{22} & -x_{12}x_{21} \end{bmatrix}.$$

Hence

$$\frac{\partial X^{-1}}{\partial X} = \frac{1}{\Delta^2}\begin{bmatrix} x_{22}^2 & -x_{12}x_{22} & -x_{22}x_{21} & x_{11}x_{22} \\ -x_{21}x_{22} & x_{12}x_{21} & x_{21}^2 & -x_{11}x_{21} \\ -x_{12}x_{22} & x_{12}^2 & x_{12}x_{21} & -x_{11}x_{12} \\ x_{11}x_{22} & -x_{12}x_{11} & -x_{11}x_{21} & x_{11}^2 \end{bmatrix}$$

$$= -\frac{1}{\Delta}\begin{bmatrix} x_{22} & -x_{12} & 0 & 0 \\ -x_{21} & x_{11} & 0 & 0 \\ 0 & 0 & x_{22} & -x_{12} \\ 0 & 0 & -x_{21} & x_{11} \end{bmatrix}\begin{bmatrix} 1 & 0 & 0 & 1 \\ 0 & 0 & 0 & 0 \\ 0 & 0 & 0 & 0 \\ 1 & 0 & 0 & 1 \end{bmatrix}\begin{bmatrix} x_{22} & -x_{12} & 0 & 0 \\ -x_{21} & x_{11} & 0 & 0 \\ 0 & 0 & x_{22} & -x_{12} \\ 0 & 0 & -x_{21} & x_{11} \end{bmatrix}$$

$$= -(I \otimes X^{-1})\,\bar{U}\,(I \otimes X^{-1})\,.$$

(4)

$$A \otimes X^{-1} = \frac{1}{\Delta}\begin{bmatrix} a_{11}x_{22} & -a_{11}x_{12} & a_{12}x_{22} & -a_{12}x_{12} \\ -a_{11}x_{21} & a_{11}x_{11} & -a_{12}x_{21} & a_{12}x_{11} \\ a_{21}x_{22} & -a_{21}x_{12} & a_{22}x_{22} & -a_{22}x_{12} \\ -a_{21}x_{21} & a_{21}x_{11} & -a_{22}x_{21} & a_{22}x_{11} \end{bmatrix}$$

where $\Delta = x_{11}x_{22} - x_{12}x_{21}$.

We can now calculate

$$\frac{\partial(A \otimes X^{-1})}{\partial x_{rs}}$$

and form

$$\frac{\partial(A \otimes X^{-1})}{\partial X} = \frac{1}{\Delta}\begin{bmatrix} 0 & 0 & 0 & 0 & 0 & -a_{11} & 0 & -a_{12} \\ 0 & a_{11} & 0 & a_{12} & 0 & 0 & 0 & 0 \\ 0 & 0 & 0 & 0 & 0 & a_{21} & 0 & -a_{22} \\ 0 & a_{21} & 0 & a_{22} & 0 & 0 & 0 & 0 \\ 0 & 0 & 0 & 0 & a_{11} & 0 & a_{12} & 0 \\ -a_{11} & 0 & -a_{12} & 0 & 0 & 0 & 0 & 0 \\ 0 & 0 & 0 & 0 & a_{21} & 0 & a_{22} & 0 \\ -a_{21} & 0 & -a_{22} & 0 & 0 & 0 & 0 & 0 \end{bmatrix}$$

Tables of Formulae and Derivatives

Table 1

Notation used: $A = [a_{ij}], B = [b_{ij}]$

$$
\begin{aligned}
E_{ij} &= \mathbf{e}_i \mathbf{e}_j' \\
\delta_{ij} &= \mathbf{e}_i' \mathbf{e}_j = \mathbf{e}_j' \mathbf{e}_i \\
E_{ij}\mathbf{e}_r &= \delta_{jr}\mathbf{e}_i \\
E_{ij}E_{rs} &= \delta_{jr}E_{is} \\
E_{ij}E_{js}E_{sm} &= E_{im} \\
E_{ij}E_{rs} &= 0 \text{ if } j \neq r \\
A &= \sum_i \sum_j a_{ij}E_{ij} \\
A_{\cdot j} &= A\mathbf{e}_j \\
A_{j\cdot} &= A'\mathbf{e}_j \\
E_{ij}AE_{rs} &= a_{jr}E_{is} \\
\operatorname{tr} AB &= \sum_i \sum_j a_{ij}b_{ji} \\
\operatorname{tr} AB' &= \operatorname{tr} A'B. \\
\operatorname{tr} AB &= (\operatorname{vec} A')' \operatorname{vec} B.
\end{aligned}
$$

Table 2

$$
\begin{aligned}
A \otimes B &= [a_{ij}B] \\
A \otimes (\alpha B) &= \alpha(A \otimes B) \\
(A + B) \otimes C &= A \otimes C + B \otimes C \\
A \otimes (B + C) &= A \otimes B + A \otimes C \\
A \otimes (B \otimes C) &= (A \otimes B) \otimes C \\
(A \otimes B)' &= A' \otimes B' \\
(A \otimes B)(C \otimes D) &= AC \otimes BC \\
(A \otimes B)^{-1} &= A^{-1} \otimes B^{-1} \\
\text{vec}\,(AYB) &= (B' \otimes A)\,\text{vec}\,Y \\
|A \otimes B| &= |A|^m |B|^n \text{ when } A \text{ and } B \text{ are of order} \\
&\quad (n \times n) \text{ and } (m \times m) \text{ respectively} \\
A \otimes B &= U_1(B \otimes A)U_2,\ U_1 \text{ and } U_2 \text{ are permutation} \\
&\quad \text{matrices} \\
\text{tr}\,(A \otimes B) &= \text{tr}\,A\ \text{tr}\,B \\
A \oplus B &= A \otimes I_m + I_n \otimes B \\
U &= \sum_r \sum_s E_{rs} \otimes E'_{rs}
\end{aligned}
$$

Table 3

$$
\begin{aligned}
\frac{\partial(A\mathbf{x})}{\partial \mathbf{x}} &= A' \\
\frac{\partial(\mathbf{x}'A)}{\partial \mathbf{x}} &= A \\
\frac{\partial(\mathbf{x}'\mathbf{x})}{\partial \mathbf{x}} &= 2\mathbf{x} \\
\frac{\partial(\mathbf{x}'A\mathbf{x})}{\partial \mathbf{x}} &= A\mathbf{x} + A'\mathbf{x} \\
\frac{\partial \mathbf{z}}{\partial \mathbf{x}} &= \frac{\partial \mathbf{y}}{\partial \mathbf{x}} \frac{\partial \mathbf{z}}{\partial \mathbf{y}}
\end{aligned}
$$

Table 4

$$\frac{\partial f(X)}{\partial X} = \Sigma\Sigma E_{ij}\frac{\partial f(X)}{\partial x_{ij}}$$

$$\frac{\partial |X|}{\partial X} = |X|(X^{-1}), \text{ when elements of } X \text{ are independent}$$

$$= 2[X_{ij}] - \text{diag}\,\{X_{ii}\}, \text{ when } X \text{ is symmetric.}$$

$$\frac{\partial X}{\partial x_{rs}} = E_{rs}$$

$$\frac{\partial X'}{\partial x_{rs}} = E'_{rs}$$

$$\frac{\partial (AXB)}{\partial x_{rs}} = AE_{rs}B$$

$$\frac{\partial (AX'B)}{\partial x_{rs}} = AE_{rs}'B$$

$$\frac{\partial (X'A'AX)}{\partial x_{rs}} = E_{rs}'A'AX + X'A'AE_{rs}$$

$$\frac{\partial (AX^{-1}B)}{\partial x_{rs}} = -AX^{-1}E_{rs}X^{-1}B$$

$$\frac{\partial (X'AX)}{\partial x_{rs}} = E'_{rs}AX + X'AE_{rs}$$

$$\frac{\partial (X^{n})}{\partial x_{rs}} = \sum_{k=0}^{n-1} X^{k}E_{rs}X^{n-k-1}$$

$$\frac{\partial (X^{-n})}{\partial x_{rs}} = -X^{-n}\left[\sum_{k=0}^{n-1} X^{k}E_{rs}X^{n-k-1}\;X^{-n}\right]$$

Table 5

$$\frac{\partial \operatorname{vec}(AXB)}{\partial \operatorname{vec} X} = B' \otimes A$$

$$\frac{\partial \operatorname{vec}(X'AX)}{\partial \operatorname{vec} X'} = U'(AX \otimes I) + (I \otimes A'X)$$

$$\frac{\partial \operatorname{vec}(AX^{-1}B)}{\partial \operatorname{vec} X} = -(X^{-1}B) \otimes (X^{-1})'A'$$

Table 6

$$\frac{\partial \log |X|}{\partial X} = (X^{-1})'$$

$$\frac{\partial |X|^r}{\partial X} = r|X|^r(X^{-1})'$$

$$\frac{\partial \operatorname{tr}(AX)}{\partial X} = A'$$

$$\frac{\partial \operatorname{tr}(A'X)}{\partial X} = A$$

$$\frac{\partial \operatorname{tr}(X'AXB)}{\partial X} = AXB + A'XB'$$

$$\frac{\partial \operatorname{tr}(XX')}{\partial X} = 2X$$

$$\frac{\partial \operatorname{tr}(X^n)}{\partial X} = nX^{n-1}$$

$$\frac{\partial \operatorname{tr}(e^X)}{\partial X} = e^X$$

$$\partial \operatorname{tr}(AX^{-1}B) = -(X^{-1}BAX^{-1})'$$

Table 7

$$\frac{\partial Y}{\partial X} = \Sigma\Sigma E_{rs} \otimes \frac{\partial Y}{\partial x_{rs}}$$

$$\frac{\partial X}{\partial X} = \bar{U} + U - \Sigma E_{rr} \otimes E_{rr} \quad (X \text{ symmetric})$$

$$\frac{\partial X}{\partial X} = \bar{U} \quad (\text{elements of } X \text{ independent})$$

$$\frac{\partial X'}{\partial X} = U$$

$$\frac{\partial (XY)}{\partial Z} = \frac{\partial X}{\partial Z}(I \otimes Y) + (I \otimes X)\frac{\partial Y}{\partial Z}$$

$$\frac{\partial X^{-1}}{\partial X} = -(I \otimes X^{-1})\bar{U}(I \otimes X^{-1})$$

$$\frac{\partial (X \otimes Y)}{\partial Z} = \frac{\partial X}{\partial Z} \otimes Y + [I \otimes U_1]\left[\frac{\partial Y}{\partial Z} \otimes X\right][I \otimes U_2]$$

Bibliography

[1] Anderson, T. W., (1958), *An Introduction to Multivariate Statistical Analysis*, John Wiley.

[2] Athans, M., (1968), *The Matrix Minimum Principle*, Information and Control, **11**, 592–606.

[3] Athans, M., and Tse, E., (1967), A Direct Derivation of the Optimal Linear Filter Using the Maximum Principle, *IEEE Trans. Auto. Control*, **AC-12**, No. 6, 690–698.

[4] Athans M., and Schweppe, F. C., (1965), *Gradient Matrices and Matrix Calculations*, MIT Lincoln Lab. Tech., Note 1965-53, Lemington, Mess.

[5] Barnett, S., (1973), Matrix Differential Equations and Kronecker Products, *SIAM, J. Appl. Math.*, **24**, No. 1.

[6] Bellman, R., (1960), *Introduction to Matrix Analysis*, McGraw-Hill.

[7] Bodewig, E., (1959), *Matrix Calculus*, Amsterdam: North Holland Publishing Co.

[8] Brewer, J. W. (1978), Kronecker Products and Matrix Calculus in System Theory, *IEEE Trans. on Circuits and Systems*, **25**, No. 9, 772–781.

[9] Brewer, J. W., (1977), The Derivatives of the Exponential Matrix with respect to a Matrix, *IEEE Trans. Auto. Control*, 22, 656–657.

[10] Brewer, J. W., (1979), Derivatives of the Characteristic Polynomial Trace and Determinant with respect to a Matrix, *IEEE Trans. Auto. Control*, **24**, 787-790.

[11] Brewer, J. W., (1977), The Gradient with respect to a Symmetric Matrix, *IEEE Trans. Auto. Control*, 22, 265-267.

[12] Brewer, J. W., (1977), The Derivative of the Riccati Matrix with respect to a Matrix, *IEEE Trans. Auto. Control*, 22, No. 6, 980-983.

[13] Conlisk, J. (1969), The Equilibrium Covariance Matrix of Dynamic Econometric Models, *American Stat. Ass. Journal*, No. 64, 277-279.

[14] Deemer, W. L. and Olkin, I., (1951), The Jacobians of certain Matrix Transformations, *Biometrika*, **30**, 345-367.

[15] Dwyer, P. S. and Macphail, M. S., (1948), Symbolic Matrix Derivatives, *Ann. Math. Statist.*, **19**, 517-537.

[16] Dwyer, P. S., (1967), Some Applications of Matrix Derivatives in Multivariate Analysis, *American Statistical Ass. Journal*, June, pt 62, 607-625.

[17] Geering, H. P., (1976), On Calculating Gradient Matrices, *IEEE Trans. Auto. Control*, August, 615-616.

[18] Graham, A., (1979), *Matrix Theory and Applications for Engineers and Mathematicians*, Ellis Horwood.

[19] Graham, A., and Burghes, D., (1980), *Introduction to Control Theory Including Optimal Control*, Ellis Horwood.

[20] Lancaster, P., (1970), Explicit Solutions of Linear Matrix Equations, *SIAM Rev.*, **12**, No. 4, 544-566.

[21] MacDuffee, C. C. (1956), *The Theory of Matrices*, Chelsea, New York.

[22] Neudecker, H. (1969), Some Theorems on Matrix Differentiation with special reference to Kronecker Matrix Products, *J. Amer. Statist. Assoc.*, **64**, 953-963.

[23] Neudecker, H., *A Note of Kronecker Matrix Products and Matrix Equation Systems.*

[24] Paraskevpoulos, P. N. and King, R. E., (1976), A Kronecker Product approach to Pole assignment by output feedback, *Int. J. Contr.*, **24**, No. 3, 325-334.

[25] Roth, W. E., (1944), On Direct Product Matrices, *Bull. Amer. Math. Soc.*, No. 40, 461-468.

[26] Schonemann, P. H., (1965), *On the Formal Differentiation of Traces and Determinants*, Research Memorandum No. 27, University of North Carolina.

[27] Schweppe, F. C., (1973), *Uncertain Dynamic Systems*, Englewood Cliffs, Prentice Hall.

[28] Tracy, D. S. and Dwyer, P. S., (1969), Multivariate Maxima and Minima with Matrix Derivatives, *J. Amer. Statist. Assoc.*, **64**, 1576-1594.

[29] Turnbull, H. W., (1927), On Differentiating a Matrix, *Proc. Edinburgh Math. Soc.*, **11**, ser. 2, 111-128.

[30] Turnbull, H. W., (1930/31), 'A Matrix Form of Taylor's Theorem', *Proc. Edinburgh Math. Soc.*, Ser. 2, 33-54.

[31] Vetter, W. J., (1970), Derivative Operations on Matrices, *IEEE Trans. Auto. Control*, **AC-15**, 241-244.

[32] Vetter, W. J., (1971), Correction to 'Derivative Operations on Matrices', *IEEE Trans. Auto. Control*, **AC-16**, 113.

[33] Vetter, W. J., (1971), An Extension to Gradient Matrices, *IEEE Trans. Syst. Man. Cybernetics*, **SMC-1**, 184-186.

[34] Vetter, W. J., (1973), Matrix Calculus Operations and Taylor Expansions, *SIAM Rev.*, **2**, 352-369.

[35] Vetter, W. J., (1975), Vector Structures and Solutions of Linear Matrix Equations, *Linear Algebra and its Applications*, **10**, 181-188.

[36] Vetter, W. J., (1971), On Linear Estimates, Minimum Variance and Least-Squares Weighting Matrices, *IEEE Trans. Auto. Control*, **AC-16**, 265-266.

[37] Weidner, R. J. and Mulholland, R. J., (1980), Kronecker Product Representation for the Solution of the General Linear Matrix Equation, *IEEE Trans. Auto. Control*, AC-25, No. 3, 563-564.

Index

A CATALOG OF SELECTED

DOVER BOOKS

IN SCIENCE AND MATHEMATICS

Astronomy

CHARIOTS FOR APOLLO: The NASA History of Manned Lunar Spacecraft to 1969, Courtney G. Brooks, James M. Grimwood, and Loyd S. Swenson, Jr. This illustrated history by a trio of experts is the definitive reference on the Apollo spacecraft and lunar modules. It traces the vehicles' design, development, and operation in space. More than 100 photographs and illustrations. 576pp. 6 3/4 x 9 1/4. 0-486-46756-2

EXPLORING THE MOON THROUGH BINOCULARS AND SMALL TELESCOPES, Ernest H. Cherrington, Jr. Informative, profusely illustrated guide to locating and identifying craters, rills, seas, mountains, other lunar features. Newly revised and updated with special section of new photos. Over 100 photos and diagrams. 240pp. 8 1/4 x 11. 0-486-24491-1

WHERE NO MAN HAS GONE BEFORE: A History of NASA's Apollo Lunar Expeditions, William David Compton. Introduction by Paul Dickson. This official NASA history traces behind-the-scenes conflicts and cooperation between scientists and engineers. The first half concerns preparations for the Moon landings, and the second half documents the flights that followed Apollo 11. 1989 edition. 432pp. 7 x 10. 0-486-47888-2

APOLLO EXPEDITIONS TO THE MOON: The NASA History, Edited by Edgar M. Cortright. Official NASA publication marks the 40th anniversary of the first lunar landing and features essays by project participants recalling engineering and administrative challenges. Accessible, jargon-free accounts, highlighted by numerous illustrations. 336pp. 8 3/8 x 10 7/8. 0-486-47175-6

ON MARS: Exploration of the Red Planet, 1958-1978--The NASA History, Edward Clinton Ezell and Linda Neuman Ezell. NASA's official history chronicles the start of our explorations of our planetary neighbor. It recounts cooperation among government, industry, and academia, and it features dozens of photos from Viking cameras. 560pp. 6 3/4 x 9 1/4. 0-486-46757-0

ARISTARCHUS OF SAMOS: The Ancient Copernicus, Sir Thomas Heath. Heath's history of astronomy ranges from Homer and Hesiod to Aristarchus and includes quotes from numerous thinkers, compilers, and scholasticists from Thales and Anaximander through Pythagoras, Plato, Aristotle, and Heraclides. 34 figures. 448pp. 5 3/8 x 8 1/2. 0-486-43886-4

AN INTRODUCTION TO CELESTIAL MECHANICS, Forest Ray Moulton. Classic text still unsurpassed in presentation of fundamental principles. Covers rectilinear motion, central forces, problems of two and three bodies, much more. Includes over 200 problems, some with answers. 437pp. 5 3/8 x 8 1/2. 0-486-64687-4

BEYOND THE ATMOSPHERE: Early Years of Space Science, Homer E. Newell. This exciting survey is the work of a top NASA administrator who chronicles technological advances, the relationship of space science to general science, and the space program's social, political, and economic contexts. 528pp. 6 3/4 x 9 1/4. 0-486-47464-X

STAR LORE: Myths, Legends, and Facts, William Tyler Olcott. Captivating retellings of the origins and histories of ancient star groups include Pegasus, Ursa Major, Pleiades, signs of the zodiac, and other constellations. "Classic." – *Sky & Telescope.* 58 illustrations. 544pp. 5 3/8 x 8 1/2. 0-486-43581-4

A COMPLETE MANUAL OF AMATEUR ASTRONOMY: Tools and Techniques for Astronomical Observations, P. Clay Sherrod with Thomas L. Koed. Concise, highly readable book discusses the selection, set-up, and maintenance of a telescope; amateur studies of the sun; lunar topography and occultations; and more. 124 figures. 26 halftones. 37 tables. 335pp. 6 1/2 x 9 1/4. 0-486-42820-6

Chemistry

MOLECULAR COLLISION THEORY, M. S. Child. This high-level monograph offers an analytical treatment of classical scattering by a central force, quantum scattering by a central force, elastic scattering phase shifts, and semi-classical elastic scattering. 1974 edition. 310pp. 5 3/8 x 8 1/2. 0-486-69437-2

HANDBOOK OF COMPUTATIONAL QUANTUM CHEMISTRY, David B. Cook. This comprehensive text provides upper-level undergraduates and graduate students with an accessible introduction to the implementation of quantum ideas in molecular modeling, exploring practical applications alongside theoretical explanations. 1998 edition. 832pp. 5 3/8 x 8 1/2. 0-486-44307-8

RADIOACTIVE SUBSTANCES, Marie Curie. The celebrated scientist's thesis, which directly preceded her 1903 Nobel Prize, discusses establishing atomic character of radioactivity; extraction from pitchblende of polonium and radium; isolation of pure radium chloride; more. 96pp. 5 3/8 x 8 1/2. 0-486-42550-9

CHEMICAL MAGIC, Leonard A. Ford. Classic guide provides intriguing entertainment while elucidating sound scientific principles, with more than 100 unusual stunts: cold fire, dust explosions, a nylon rope trick, a disappearing beaker, much more. 128pp. 5 3/8 x 8 1/2. 0-486-67628-5

ALCHEMY, E. J. Holmyard. Classic study by noted authority covers 2,000 years of alchemical history: religious, mystical overtones; apparatus; signs, symbols, and secret terms; advent of scientific method, much more. Illustrated. 320pp. 5 3/8 x 8 1/2. 0-486-26298-7

CHEMICAL KINETICS AND REACTION DYNAMICS, Paul L. Houston. This text teaches the principles underlying modern chemical kinetics in a clear, direct fashion, using several examples to enhance basic understanding. Solutions to selected problems. 2001 edition. 352pp. 8 3/8 x 11. 0-486-45334-0

PROBLEMS AND SOLUTIONS IN QUANTUM CHEMISTRY AND PHYSICS, Charles S. Johnson and Lee G. Pedersen. Unusually varied problems, with detailed solutions, cover of quantum mechanics, wave mechanics, angular momentum, molecular spectroscopy, scattering theory, more. 280 problems, plus 139 supplementary exercises. 430pp. 6 1/2 x 9 1/4. 0-486-65236-X

ELEMENTS OF CHEMISTRY, Antoine Lavoisier. Monumental classic by the founder of modern chemistry features first explicit statement of law of conservation of matter in chemical change, and more. Facsimile reprint of original (1790) Kerr translation. 539pp. 5 3/8 x 8 1/2. 0-486-64624-6

MAGNETISM AND TRANSITION METAL COMPLEXES, F. E. Mabbs and D. J. Machin. A detailed view of the calculation methods involved in the magnetic properties of transition metal complexes, this volume offers sufficient background for original work in the field. 1973 edition. 240pp. 5 3/8 x 8 1/2. 0-486-46284-6

GENERAL CHEMISTRY, Linus Pauling. Revised third edition of classic first-year text by Nobel laureate. Atomic and molecular structure, quantum mechanics, statistical mechanics, thermodynamics correlated with descriptive chemistry. Problems. 992pp. 5 3/8 x 8 1/2. 0-486-65622-5

ELECTROLYTE SOLUTIONS: Second Revised Edition, R. A. Robinson and R. H. Stokes. Classic text deals primarily with measurement, interpretation of conductance, chemical potential, and diffusion in electrolyte solutions. Detailed theoretical interpretations, plus extensive tables of thermodynamic and transport properties. 1970 edition. 590pp. 5 3/8 x 8 1/2. 0-486-42225-9

Engineering

FUNDAMENTALS OF ASTRODYNAMICS, Roger R. Bate, Donald D. Mueller, and Jerry E. White. Teaching text developed by U.S. Air Force Academy develops the basic two-body and n-body equations of motion; orbit determination; classical orbital elements, coordinate transformations; differential correction; more. 1971 edition. 455pp. 5 3/8 x 8 1/2. 0-486-60061-0

INTRODUCTION TO CONTINUUM MECHANICS FOR ENGINEERS: Revised Edition, Ray M. Bowen. This self-contained text introduces classical continuum models within a modern framework. Its numerous exercises illustrate the governing principles, linearizations, and other approximations that constitute classical continuum models. 2007 edition. 320pp. 6 1/8 x 9 1/4. 0-486-47460-7

ENGINEERING MECHANICS FOR STRUCTURES, Louis L. Bucciarelli. This text explores the mechanics of solids and statics as well as the strength of materials and elasticity theory. Its many design exercises encourage creative initiative and systems thinking. 2009 edition. 320pp. 6 1/8 x 9 1/4. 0-486-46855-0

FEEDBACK CONTROL THEORY, John C. Doyle, Bruce A. Francis and Allen R. Tannenbaum. This excellent introduction to feedback control system design offers a theoretical approach that captures the essential issues and can be applied to a wide range of practical problems. 1992 edition. 224pp. 6 1/2 x 9 1/4. 0-486-46933-6

THE FORCES OF MATTER, Michael Faraday. These lectures by a famous inventor offer an easy-to-understand introduction to the interactions of the universe's physical forces. Six essays explore gravitation, cohesion, chemical affinity, heat, magnetism, and electricity. 1993 edition. 96pp. 5 3/8 x 8 1/2. 0-486-47482-8

DYNAMICS, Lawrence E. Goodman and William H. Warner. Beginning engineering text introduces calculus of vectors, particle motion, dynamics of particle systems and plane rigid bodies, technical applications in plane motions, and more. Exercises and answers in every chapter. 619pp. 5 3/8 x 8 1/2. 0-486-42006-X

ADAPTIVE FILTERING PREDICTION AND CONTROL, Graham C. Goodwin and Kwai Sang Sin. This unified survey focuses on linear discrete-time systems and explores natural extensions to nonlinear systems. It emphasizes discrete-time systems, summarizing theoretical and practical aspects of a large class of adaptive algorithms. 1984 edition. 560pp. 6 1/2 x 9 1/4. 0-486-46932-8

INDUCTANCE CALCULATIONS, Frederick W. Grover. This authoritative reference enables the design of virtually every type of inductor. It features a single simple formula for each type of inductor, together with tables containing essential numerical factors. 1946 edition. 304pp. 5 3/8 x 8 1/2. 0-486-47440-2

THERMODYNAMICS: Foundations and Applications, Elias P. Gyftopoulos and Gian Paolo Beretta. Designed by two MIT professors, this authoritative text discusses basic concepts and applications in detail, emphasizing generality, definitions, and logical consistency. More than 300 solved problems cover realistic energy systems and processes. 800pp. 6 1/8 x 9 1/4. 0-486-43932-1

THE FINITE ELEMENT METHOD: Linear Static and Dynamic Finite Element Analysis, Thomas J. R. Hughes. Text for students without in-depth mathematical training, this text includes a comprehensive presentation and analysis of algorithms of time-dependent phenomena plus beam, plate, and shell theories. Solution guide available upon request. 672pp. 6 1/2 x 9 1/4. 0-486-41181-8

Mathematics-Bestsellers

HANDBOOK OF MATHEMATICAL FUNCTIONS: with Formulas, Graphs, and Mathematical Tables, Edited by Milton Abramowitz and Irene A. Stegun. A classic resource for working with special functions, standard trig, and exponential logarithmic definitions and extensions, it features 29 sets of tables, some to as high as 20 places. 1046pp. 8 x 10 1/2. 0-486-61272-4

ABSTRACT AND CONCRETE CATEGORIES: The Joy of Cats, Jiri Adamek, Horst Herrlich, and George E. Strecker. This up-to-date introductory treatment employs category theory to explore the theory of structures. Its unique approach stresses concrete categories and presents a systematic view of factorization structures. Numerous examples. 1990 edition, updated 2004. 528pp. 6 1/8 x 9 1/4. 0-486-46934-4

MATHEMATICS: Its Content, Methods and Meaning, A. D. Aleksandrov, A. N. Kolmogorov, and M. A. Lavrent'ev. Major survey offers comprehensive, coherent discussions of analytic geometry, algebra, differential equations, calculus of variations, functions of a complex variable, prime numbers, linear and non-Euclidean geometry, topology, functional analysis, more. 1963 edition. 1120pp. 5 3/8 x 8 1/2. 0-486-40916-3

INTRODUCTION TO VECTORS AND TENSORS: Second Edition--Two Volumes Bound as One, Ray M. Bowen and C.-C. Wang. Convenient single-volume compilation of two texts offers both introduction and in-depth survey. Geared toward engineering and science students rather than mathematicians, it focuses on physics and engineering applications. 1976 edition. 560pp. 6 1/2 x 9 1/4. 0-486-46914-X

AN INTRODUCTION TO ORTHOGONAL POLYNOMIALS, Theodore S. Chihara. Concise introduction covers general elementary theory, including the representation theorem and distribution functions, continued fractions and chain sequences, the recurrence formula, special functions, and some specific systems. 1978 edition. 272pp. 5 3/8 x 8 1/2. 0-486-47929-3

ADVANCED MATHEMATICS FOR ENGINEERS AND SCIENTISTS, Paul DuChateau. This primary text and supplemental reference focuses on linear algebra, calculus, and ordinary differential equations. Additional topics include partial differential equations and approximation methods. Includes solved problems. 1992 edition. 400pp. 7 1/2 x 9 1/4. 0-486-47930-7

PARTIAL DIFFERENTIAL EQUATIONS FOR SCIENTISTS AND ENGINEERS, Stanley J. Farlow. Practical text shows how to formulate and solve partial differential equations. Coverage of diffusion-type problems, hyperbolic-type problems, elliptic-type problems, numerical and approximate methods. Solution guide available upon request. 1982 edition. 414pp. 6 1/8 x 9 1/4. 0-486-67620-X

VARIATIONAL PRINCIPLES AND FREE-BOUNDARY PROBLEMS, Avner Friedman. Advanced graduate-level text examines variational methods in partial differential equations and illustrates their applications to free-boundary problems. Features detailed statements of standard theory of elliptic and parabolic operators. 1982 edition. 720pp. 6 1/8 x 9 1/4. 0-486-47853-X

LINEAR ANALYSIS AND REPRESENTATION THEORY, Steven A. Gaal. Unified treatment covers topics from the theory of operators and operator algebras on Hilbert spaces; integration and representation theory for topological groups; and the theory of Lie algebras, Lie groups, and transform groups. 1973 edition. 704pp. 6 1/8 x 9 1/4. 0-486-47851-3

Mathematics–Logic and Problem Solving

Mathematics–Algebra and Calculus

VECTOR CALCULUS, Peter Baxandall and Hans Liebeck. This introductory text offers a rigorous, comprehensive treatment. Classical theorems of vector calculus are amply illustrated with figures, worked examples, physical applications, and exercises with hints and answers. 1986 edition. 560pp. 5 3/8 x 8 1/2. 0-486-46620-5

ADVANCED CALCULUS: An Introduction to Classical Analysis, Louis Brand. A course in analysis that focuses on the functions of a real variable, this text introduces the basic concepts in their simplest setting and illustrates its teachings with numerous examples, theorems, and proofs. 1955 edition. 592pp. 5 3/8 x 8 1/2. 0-486-44548-8

ADVANCED CALCULUS, Avner Friedman. Intended for students who have already completed a one-year course in elementary calculus, this two-part treatment advances from functions of one variable to those of several variables. Solutions. 1971 edition. 432pp. 5 3/8 x 8 1/2. 0-486-45795-8

METHODS OF MATHEMATICS APPLIED TO CALCULUS, PROBABILITY, AND STATISTICS, Richard W. Hamming. This 4-part treatment begins with algebra and analytic geometry and proceeds to an exploration of the calculus of algebraic functions and transcendental functions and applications. 1985 edition. Includes 310 figures and 18 tables. 880pp. 6 1/2 x 9 1/4. 0-486-43945-3

BASIC ALGEBRA I: Second Edition, Nathan Jacobson. A classic text and standard reference for a generation, this volume covers all undergraduate algebra topics, including groups, rings, modules, Galois theory, polynomials, linear algebra, and associative algebra. 1985 edition. 528pp. 6 1/8 x 9 1/4. 0-486-47189-6

BASIC ALGEBRA II: Second Edition, Nathan Jacobson. This classic text and standard reference comprises all subjects of a first-year graduate-level course, including in-depth coverage of groups and polynomials and extensive use of categories and functors. 1989 edition. 704pp. 6 1/8 x 9 1/4. 0-486-47187-X

CALCULUS: An Intuitive and Physical Approach (Second Edition), Morris Kline. Application-oriented introduction relates the subject as closely as possible to science with explorations of the derivative; differentiation and integration of the powers of x; theorems on differentiation, antidifferentiation; the chain rule; trigonometric functions; more. Examples. 1967 edition. 960pp. 6 1/2 x 9 1/4. 0-486-40453-6

ABSTRACT ALGEBRA AND SOLUTION BY RADICALS, John E. Maxfield and Margaret W. Maxfield. Accessible advanced undergraduate-level text starts with groups, rings, fields, and polynomials and advances to Galois theory, radicals and roots of unity, and solution by radicals. Numerous examples, illustrations, exercises, appendixes. 1971 edition. 224pp. 6 1/8 x 9 1/4. 0-486-47723-1

AN INTRODUCTION TO THE THEORY OF LINEAR SPACES, Georgi E. Shilov. Translated by Richard A. Silverman. Introductory treatment offers a clear exposition of algebra, geometry, and analysis as parts of an integrated whole rather than separate subjects. Numerous examples illustrate many different fields, and problems include hints or answers. 1961 edition. 320pp. 5 3/8 x 8 1/2. 0-486-63070-6

LINEAR ALGEBRA, Georgi E. Shilov. Covers determinants, linear spaces, systems of linear equations, linear functions of a vector argument, coordinate transformations, the canonical form of the matrix of a linear operator, bilinear and quadratic forms, and more. 387pp. 5 3/8 x 8 1/2. 0-486-63518-X

Mathematics–Probability and Statistics

BASIC PROBABILITY THEORY, Robert B. Ash. This text emphasizes the probabilistic way of thinking, rather than measure-theoretic concepts. Geared toward advanced undergraduates and graduate students, it features solutions to some of the problems. 1970 edition. 352pp. 5 3/8 x 8 1/2. 0-486-46628-0

PRINCIPLES OF STATISTICS, M. G. Bulmer. Concise description of classical statistics, from basic dice probabilities to modern regression analysis. Equal stress on theory and applications. Moderate difficulty; only basic calculus required. Includes problems with answers. 252pp. 5 5/8 x 8 1/4. 0-486-63760-3

OUTLINE OF BASIC STATISTICS: Dictionary and Formulas, John E. Freund and Frank J. Williams. Handy guide includes a 70-page outline of essential statistical formulas covering grouped and ungrouped data, finite populations, probability, and more, plus over 1,000 clear, concise definitions of statistical terms. 1966 edition. 208pp. 5 3/8 x 8 1/2. 0-486-47769-X

GOOD THINKING: The Foundations of Probability and Its Applications, Irving J. Good. This in-depth treatment of probability theory by a famous British statistician explores Keynesian principles and surveys such topics as Bayesian rationality, corroboration, hypothesis testing, and mathematical tools for induction and simplicity. 1983 edition. 352pp. 5 3/8 x 8 1/2. 0-486-47438-0

INTRODUCTION TO PROBABILITY THEORY WITH CONTEMPORARY APPLICATIONS, Lester L. Helms. Extensive discussions and clear examples, written in plain language, expose students to the rules and methods of probability. Exercises foster problem-solving skills, and all problems feature step-by-step solutions. 1997 edition. 368pp. 6 1/2 x 9 1/4. 0-486-47418-6

CHANCE, LUCK, AND STATISTICS, Horace C. Levinson. In simple, non-technical language, this volume explores the fundamentals governing chance and applies them to sports, government, and business. "Clear and lively ... remarkably accurate." – *Scientific Monthly*. 384pp. 5 3/8 x 8 1/2. 0-486-41997-5

FIFTY CHALLENGING PROBLEMS IN PROBABILITY WITH SOLUTIONS, Frederick Mosteller. Remarkable puzzlers, graded in difficulty, illustrate elementary and advanced aspects of probability. These problems were selected for originality, general interest, or because they demonstrate valuable techniques. Also includes detailed solutions. 88pp. 5 3/8 x 8 1/2. 0-486-65355-2

EXPERIMENTAL STATISTICS, Mary Gibbons Natrella. A handbook for those seeking engineering information and quantitative data for designing, developing, constructing, and testing equipment. Covers the planning of experiments, the analyzing of extreme-value data; and more. 1966 edition. Index. Includes 52 figures and 76 tables. 560pp. 8 3/8 x 11. 0-486-43937-2

STOCHASTIC MODELING: Analysis and Simulation, Barry L. Nelson. Coherent introduction to techniques also offers a guide to the mathematical, numerical, and simulation tools of systems analysis. Includes formulation of models, analysis, and interpretation of results. 1995 edition. 336pp. 6 1/8 x 9 1/4. 0-486-47770-3

INTRODUCTION TO BIOSTATISTICS: Second Edition, Robert R. Sokal and F. James Rohlf. Suitable for undergraduates with a minimal background in mathematics, this introduction ranges from descriptive statistics to fundamental distributions and the testing of hypotheses. Includes numerous worked-out problems and examples. 1987 edition. 384pp. 6 1/8 x 9 1/4. 0-486-46961-1